UNDECIDABLE
THEORIES

UNDECIDABLE THEORIES

ALFRED TARSKI

in collaboration with

Andrzej Mostowski and

Raphael M. Robinson

DOVER PUBLICATIONS, INC.

Mineola, New York

Bibliographical Note

This Dover edition, first published in 2010, is an unabridged republication of the work originally published by North-Holland Publishing Company, Amsterdam, in 1953.

International Standard Book Number

ISBN-13: 978-0-486-47703-9
ISBN-10: 0-486-47703-7

www.doverpublications.com

To

HEINRICH SCHOLZ

THE SCHOLAR AND THE MAN

PREFACE

This monograph consists of three papers: *A general method in proofs of undecidability, Undecidability and essential undecidability in arithmetic, Undecidability of the elementary theory of groups.* While the first and the third papers have been written by the undersigned alone, the second paper is a joint work of A. Mostowski, R. M. Robinson, and the undersigned.

The three papers are referred to throughout the monograph by Roman numerals I, II, III. The introduction to paper I is thought of as an introduction to the whole work and gives an idea of the scope of the problems discussed and the results obtained. The notations and symbolic conventions introduced at any place of the monograph are applied in the whole subsequent discussion.

The work contains results obtained over a long period of time, 1938–1952. The first draft of paper II was prepared by Mostowski in 1949 and contained exclusively results found by him and Tarski in the pre-war period. However, its publication was postponed and the text was considerably modified, so as to embody the results and simplifications subsequently obtained by Robinson and partly by Tarski; the final draft was written jointly by these two authors in 1951–52. The other two papers were prepared for publication during the same period.

It was originally planned to publish the work as a series of connected papers in some regular mathematical periodical. However, the Editors of the series *Studies in Logic* found the material integrated enough to appear as a separate volume in this series; their offer was accepted with appreciation.

A large part of the technical work on the monograph was done during the period when Tarski and, for a shorter time, also Robinson were engaged in a research project in the foundations of mathematics sponsored by the Office of Ordnance Research, U.S. Army.

In this work the authors received much valuable advice and assistance from several friends—Mr. C. C. Chang, Professor Leon Henkin, Professor J. C. C. McKinsey, Dr. Julia Robinson, and Mr. R. L. Vaught. Professor E. W. Beth and Mr. F. W. J. Marx have greatly obliged the authors with their help in reading proofs.

University of California
Berkeley, April 1953

Alfred Tarski

CONTENTS

I

A GENERAL METHOD IN PROOFS OF UNDECIDABILITY

BY

ALFRED TARSKI

I

A GENERAL METHOD IN PROOFS OF UNDECIDABILITY

I.1. Introduction. [1] By a *decision procedure* for a given formalized theory T we understand a method which permits us to decide in each particular case whether a given sentence formulated in the symbolism of T can be proved by means of the devices available in T (or, more generally, can be recognized as valid in T). [2] The *decision problem* for T is the problem of determining whether a decision procedure for T exists (and possibly of exhibiting such a procedure). A theory T is called *decidable* or *undecidable* according as the solution of the decision problem is positive or negative. As is well known, the decision problem is one of the central problems of contemporary metamathematics. Since only few theories turn out to be decidable [3], most endeavors are directed toward a negative solution.

In the attacks on the decision problem for various special theories and, more specifically, in the attempts at obtaining a negative solution of this problem two different methods are applied. The first, direct method is essentially based upon ideas which originated with Gödel and were used for the first time in [7], in the proofs of his well-known incompleteness theorems. When applying this method we have to make use of some deep properties of notions which are involved in the precise statement of the decision problem; in fact, of the notions of general recursive functions and sets. The method is rather involved and can be applied only to those theories in which a sufficiently strong number-theoretical apparatus

---

[1] The observations contained in this paper were made in 1938–1939; they were presented by the author to a meeting of the Association for Symbolic Logic in 1948, and were summarized in [34]. (The numbers in square brackets refer to Bibliography.)

[2] The meanings of various terms used in this introduction will be explained in a more detailed and precise way in later sections of the paper.

[3] Some decidable theories are discussed or at least mentioned in [30].

can be developed. With the help of this method some fundamental
results concerning the decision problem have been obtained. In
fact, it was shown by Church in [3] that the solution of the decision
problem for Peano's arithmetic (and some fragments of it) is negative, and Rosser proved later in [24] that the same applies to every
consistent theory which is an extension of Peano's arithmetic.
We express these results briefly by saying that Peano's arithmetic
is, not only undecidable, but also *essentially undecidable.* — The
theoretical foundations of the direct method will be outlined in II,
in particular in II.2 and II.4.

The second, indirect method consists in reducing the decision
problem for a theory T_1 to the decision problem for some other
theory T_2 for which the problem has previously been solved. In
the original form of this method, to establish the undecidability
of a theory T_1 one tried to show that either (i) T_1 can be obtained
from some undecidable theory T_2 by deleting finitely many axioms
from the axiom system of T_2 (but without removing any constant
from the symbolism of T_2), or else that (ii) some essentially undecidable theory T_2 is interpretable in T_1. By applying the procedure
(i) and by taking a fragment of Peano's arithmetic (suitably modified) for T_2, Church proved in [2] that the first-order predicate
logic is undecidable. When applying the procedure (ii), Peano's
arithmetic is usually taken for T_2; in this way, e.g., various axiomatic
systems of set theory have turned out to be undecidable.

The indirect method in its original form was rather restricted
in applications. Only in exceptional cases can a theory for which
the decision problem is discussed be obtained from another theory,
which is known to be undecidable, simply by omitting finitely
many sentences from the axiom system of the latter. On the other
hand, one could hardly expect to find an interpretation of Peano's
arithmetic in various simple formalized theories, with meager
mathematical contents, for which the decision problem was open.
With regard to theories of this kind both the direct and the indirect
methods seemed to fail. However, it has proved to be possible to
extend and modify the indirect method (by combining some
features of the two procedures indicated above) so as to widen

considerably its range of application. In fact, a simple argument shows that, in order to establish the undecidability of a theory T_1, it suffices to show that some essentially undecidable theory T_2 can be interpreted, not necessarily in T_1, but (what is much easier) in some consistent extension of T_1 — provided only that T_2 is based upon a finite axiom system. As a consequence of this last condition, Peano's arithmetic cannot any longer be used as T_2 since it is not based upon a finite axiom system. On the other hand, examples of essentially undecidable theories which are based upon finite axiom systems and are readily interpretable in other theories have been found (by the direct method) among fragments of Peano's arithmetic. Using this fact and applying the extended indirect method, many formalized theories—like the elementary theories of groups, rings, fields, and lattices—have recently been shown to be undecidable.

The aim of the present paper is to set up theoretical foundations for the general method just described. The paper is conceived as a framework for later publications in this field, with the idea of permitting the authors to avoid the repetition of some lengthy, though elementary, discussions. With this in view, we give in I.2 a rather detailed description of formalized theories to which the method applies. In I.3–5 we define the notions involved in the description of the method, and we explicitly state and prove a few elementary theorems upon which this method is based. Finally, I.6 contains a short survey of the results obtained so far with the help of the method discussed. Throughout the whole paper the discussion has an informal character.

I.2. **Theories with standard formalization.** The theories discussed in this paper will be referred to as *theories with standard formalization*. They can be briefly characterized as theories which are formalized within the first-order predicate logic (with identity, without variable predicates). [4]

---

[4] For a detailed discussion of predicate logic and theories formalized within this logic consult, e.g., [9]; the discussion is spread over vol. 1.of this work, and is summarized and supplemented in vol. 2, pp. 375–391. The reader

The symbols which occur in expressions of a given theory T are divided into *variables* and *constants*. The set of variables is assumed to be denumerable and hence infinite; the set of constants is either finite or denumerable. All the variables are treated as ranging over the same set of elements. The constants are divided into *logical* and *non-logical* ones. The logical constants are the *sentential connectives*—the *negation sign* $\sim$, the *implication sign* $\rightarrow$, the *equivalence sign* $\leftrightarrow$, the *disjunction sign* $\vee$, and the *conjunction sign* $\wedge$; the *quantifiers*—the *universal quantifier* $\wedge$ and the *existential quantifier* $\vee$; and, finally, the *identity symbol* $=$. [5] The non-logical constants are the *predicates* (or *relation symbols*), the *operation symbols*, and the *individual constants*. With every predicate and every operation symbol a positive integer is correlated which is called the *rank* of the symbol. Thus, we may have in T *unary* predicates and operation symbols (i.e., symbols of rank 1), *binary* predicates and operation symbols (symbols of rank 2), etc. The identity symbol, though regarded as a logical constant, is included in the set of binary predicates. In practice, in addition to variables and constants, the so-called technical symbols, like parentheses and commas, are also used in constructing expressions; theoretically, however, these technical symbols can be dispensed with.

Among expressions (i.e., finite concatenations of symbols) we

---

will find there elaborations of certain points which have been disregarded in the present account (e.g., the definitions of free and bound variables, the specification of logical axioms and operations of inference, and the proofs of the deduction theorems).

[5] In view of the general character of our present discussion we have not considered it necessary to maintain a strict distinction between expressions of formalized theories and their metamathematical designations. No inconsistency will arise, however, if we agree to regard all the symbolic expressions used in this paper, not as expressions of the theories discussed, but as metamathematical denotations of such expressions. In this case the letters "x", "y", "z", ... should be regarded as metamathematical variables which range over variables of the theory under discussion, and usually it should be assumed that two different letters represent two distinct variables of the theory.

distinguish *terms* and *formulas*. The simplest, so-called atomic, terms are the variables and the individual constants; a compound term is obtained by combining n simpler terms by means of an operation symbol of rank n. Similarly, an atomic formula is obtained by combining n arbitrary terms by means of a predicate of rank n; compound formulas are built from simpler ones by means of sentential connectives and quantifier expressions (i.e., quantifiers followed by variables, like $\wedge x$ or $\vee y$). An occurrence of a variable in a formula may be either *free* or *bound*; a formula in which no variable occurs free is called a *sentence*.

Two further notions, those of logical derivability and logical validity, are involved in the metamathematical discussion of any theory T. They are usually introduced in the following way. First, we single out certain sentences of T which are referred to as *logical axioms*. Secondly, we describe certain (finitary) operations, the so-called *operations of inference*, which when performed on sentences yield new sentences. Usually the set of logical axioms is infinite while the set of operations of inference is finite. The most important operation of inference is that of *detachment* (*modus ponens*), which when applied to two sentences Φ and $\Phi \to \Psi$ yields the sentence Ψ. In fact, it proves to be possible, by selecting a suitable set of logical axioms, to use the operation of detachment as the only operation of inference in formulating adequate definitions of derivability and logical validity. [6]

A sentence is now said to be *logically derivable* or simply *derivable* from a set A of sentences if it can be obtained from sentences of A and from logical axioms by performing operations of inference an arbitrary number of times. A sentence is called *logically valid* (or *logically provable*) if it is derivable from the set of logical axioms—or, what amounts to the same thing, from the empty set of sentences.

Still another method of defining these two notions is available which, however, essentially involves the use of some semantical

---

[6] Cf. in this connection the remarks in [33], pp. 507 f. A suitable system of logical axioms is given in [19], pp. 80–85; the system is to be supplemented by axioms involving the identity symbol.

notions and, in particular, of the notion of satisfaction. [7] We assume that all the non-logical constants of T have been arranged in a (finite or infinite) sequence $\langle \mathbf{C}_0, \ldots, \mathbf{C}_n, \ldots \rangle$, without repeating terms. We consider systems $\mathfrak{R}$ formed by a non-empty set U and by a sequence $\langle C_0, \ldots, C_n, \ldots \rangle$ of certain mathematical entities, with the same number of terms as the sequence of non-logical constants. The mathematical nature of each C_n depends on the logical character of the corresponding constant $\mathbf{C}_n$. Thus, if $\mathbf{C}_n$ is a unary predicate, then C_n is a subset of U; more generally, if $\mathbf{C}_n$ is an m-ary predicate, then C_n is an m-ary relation the field of which is a subset of U. If $\mathbf{C}_n$ is an m-ary operation symbol, C_n is an m-ary operation (function of m arguments) defined over arbitrary ordered m-tuples $\langle x_1, \ldots, x_m \rangle$ of elements of U and assuming elements of U as values. If, finally, $\mathbf{C}_n$ is an individual constant, C_n is simply an element of U. Such a system (sequence) $\mathfrak{R} = \langle U, C_0, \ldots, C_n, \ldots \rangle$ is called a *possible realization* or simply a *realization* of T; the set U is called the *universe* of $\mathfrak{R}$. We assume it to be clear under what conditions a sentence Φ of T is said to *be satisfied* or to *hold* in a given realization $\mathfrak{R}$. Roughly speaking, this means that Φ turns out to be true if (i) all the variables occurring in T are assumed to range over the set U; (ii) the logical constants are interpreted in the usual way; (iii) each of the non-logical constants $\mathbf{C}_n$ is understood to denote the corresponding term C_n in $\mathfrak{R}$. (Assume, e.g., that the term $\mathbf{C}_n$ in the sequence of constants is a unary predicate and that consequently C_n is a subset of U. Then the sentence $\wedge x \mathbf{C}_n x$ holds in $\mathfrak{R}$ if and only if every element of U is an element of C_n and hence C_n coincides with U.) A sentence Φ is said to be a *logical consequence* of a set A of sentences if it is satisfied in every realization $\mathfrak{R}$ in which all sentences of A are satisfied; it is called *logically true* if it is satisfied in every possible realization.

As opposed to the notions of logical consequence and logical truth, the related notions of logical derivability and logical validity,

---

[7] For formal definitions and a detailed discussion of semantical notions (satisfaction, truth, logical consequence, logical truth) see [31] and [38].

when defined in terms of axioms and operations of inference, seem to have a rather accidental and arbitrary character. Hence it might seem natural to redefine these notions simply by stipulating that a sentence is derivable from A if it is a logical consequence of A, and by identifying logically valid sentences with logically true sentences. From the results in Gödel [5] it follows, however, that under the systems of logical axioms and operations of inference known from the literature the two methods of defining derivability and logical validity are entirely equivalent (when applied to theories with standard formalization).

An important property of the notion of derivability is stated in the following well-known theorem, which is often applied in metamathematical discussion:

DEDUCTION THEOREM I. *Let A be a set of sentences of a theory T and let $\Phi_1, \ldots, \Phi_n, \Psi$ be any sentences of T. For Ψ to be derivable from the set A supplemented by the sentences $\Phi_1, \ldots, \Phi_n$ it is necessary and sufficient that the sentence*

$$(\Phi_1 \wedge \ldots \wedge \Phi_n) \to \Psi$$

be derivable from the set A alone.

With the help of this theorem we obtain

DEDUCTION THEOREM II. *Let A be a set of sentences of a theory T, and let Ψ be a sentence of T. For Ψ to be derivable from A it is necessary and sufficient that A be empty and Ψ be logically valid or else that A contain some sentences $\Phi_1, \ldots, \Phi_n$ such that the sentence*

$$(\Phi_1 \wedge \ldots \wedge \Phi_n) \to \Psi$$

is logically valid.

Thus the notion of derivability has a simple characterization in terms of logically valid sentences. [8]

---

[8] The proofs of the deduction theorems are very simple under the assumptions that the operation of detachment is used as the only operation of inference and that all the tautological sentences (in the sense of sentential

To complete the description of a theory T we have to define what we mean by a *valid* sentence in general (as opposed to a logically valid sentence). No uniform method for defining this notion is available. Often we single out a (finite or infinite) set of sentences called *non-logical axioms,* and define a sentence to be valid if and only if it is derivable from this set—or, what amounts to the same, from the set of all axioms, both logical and non-logical. [9] Theories in which the notion of validity has been introduced in this way are referred to as *axiomatically built* or, simply, *axiomatic* theories; when referring to such theories, we often use the term *"provable"* instead of *"valid".* We do not restrict ourselves to the discussion of axiomatic theories. Sometimes we agree to consider as valid those and only those sentences which are satisfied in a given realization or in all realizations of a given class; sometimes we define validity for a theory T in terms of validity for some other theories for which this notion has been previously defined. We assume, however, that for each of the theories discussed

---

calculus) are known to be logically valid. Under the same assumptions we can establish a further result closely related to Deduction Theorem II; in fact, we can show that a sentence Ψ is derivable from a set A if and only if there are sentences $\Phi_1, \ldots, \Phi_n$ which are logical axioms or members of A and for which the sentence

$$(\Phi_1 \wedge \ldots \wedge \Phi_n) \to \Psi$$

is tautological. Hence, by including all tautological sentences in the set of logical axioms, we obtain a simple characterization of derivability in terms of logical axioms.

It may be mentioned that Deduction Theorem II has sometimes been used as a definition of the notion of derivability. See [33], p. 507, and [38], pp. 10–11, footnote 10 (where references to earlier papers of Ajdukiewicz can also be found).

[9] We can of course transform this definition by means of Deduction Theorem II so as to obtain a characterization of valid sentences in terms of logically valid sentences and non-logical axioms. Under conditions stated in footnote 8 a further simplification is possible; we obtain a simple characterization of valid sentences in terms of arbitrary (logical and non-logical) axioms, a characterization in which the notions of derivability and logical validity do not appear and no repeated application of operations of inference is involved.

the notion of validity has been defined in one way or another. We also assume that under this definition every sentence which is logically derivable from a set of valid sentences is itself valid, and that consequently every logically valid sentence is valid; this is the only condition imposed upon the definition of validity. A possible realization in which all valid sentences of a theory T are satisfied is called a *model* of T.

In all the theories with standard formalization the same symbols are assumed to be used as variables and logical constants; apart from differences in non-logical constants, the same expressions are regarded as formulas, sentences, logical axioms, and logically valid sentences. However, the notions of validity in these theories may of course exhibit essential differences. A theory is uniquely determined by the set of all its valid sentences; two theories are regarded as identical if their sets of valid sentences coincide. An axiomatic theory is uniquely determined by its non-logical constants and non-logical axioms.

A theory T_1 is called a *subtheory* of a theory T_2 if every sentence which is valid in T_1 is also valid in T_2; under the same conditions T_2 is referred to as an *extension* of T_1. An extension T_2 of T_1 is called *inessential* if every constant of T_2 which does not occur in T_1 is an individual constant and if every valid sentence of T_2 is derivable in T_2 from a set of valid sentences of T_1. (By saying that a sentence Φ is derivable *in a theory* T from a set A we stress the fact that, in deriving Φ, we may use both sentences of A and logical axioms of T. It is easily seen that, whenever Φ is derivable from A in some theory T, it is also derivable from A in every theory T' which contains all the non-logical constants occurring in Φ and in sentences of A.) If T_1 is axiomatic, then an inessential extension of T_1 is obtained by adding some new individual constants, but without adding any new non-logical axioms. An extension T_2 of T_1 is referred to as a *finite* extension if there is a finite set A of valid sentences of T_2 such that every valid sentence of T_2 is derivable from a set of sentences which are valid in T_1 or belong to A. Clearly every inessential extension is a finite extension.

Among the extensions common to two given theories T_1 and T_2 there is always a smallest one, which is a subtheory of any other common extension; this smallest common extension is referred to as the *union* of the given theories. The union T of T_1 and T_2 is fully characterized by the following two conditions: (i) the set of all non-logical constants of T is the (set-theoretical) union of the sets of all non-logical constants of T_1 and T_2; (ii) a sentence is valid in T if and only if it is derivable in T from a set of sentences which are valid in T_1 or T_2. (Notice that condition (i) unambiguously determines the notions of a sentence of T and of a logical axiom of T, and hence also the notion of derivability in T.) If the theories T_1 and T_2 are axiomatic, we can construct T by postulating, in addition to (i), the analogous condition for the set of non-logical axioms.

A theory T is called *consistent* if not every sentence of T is valid in T; or, in an equivalent formulation, if there is no sentence Φ such that both Φ and $\sim \Phi$ are valid in T. A theory T is called *complete* if there is no consistent extension of T which is different from T, but which has the same constants as T; or, equivalently, if, for every sentence Φ of T, either Φ or $\sim \Phi$ is valid in T. The proof of the equivalence of the two definitions of completeness is based upon Deduction Theorem I. Consistency and completeness can also be characterized in terms of models: a theory T is consistent if and only if it has at least one model; it is complete if and only if every sentence of T which is satisfied in one model is also satisfied in any other model of T. Two theories T_1 and T_2 are said to be *compatible* if they have a common consistent extension; this is equivalent to saying that the union of T_1 and T_2 is consistent.

I.3. **Undecidable and essentially undecidable theories.** Since the set of all expressions of a given theory T is denumerable, it is possible to establish a one-to-one correspondence between expressions of T and natural numbers. Having fixed such a correspondence, we can extend to expressions various notions originally defined for natural numbers, and conversely. [10] In particular, a set E of

---

[10] This was noticed at about the same time, but independently, by Gödel and the author. Cf. [7] where the arithmetization of metamathematics has

expressions is said to be *general recursive* if the set N of natural numbers correlated with expressions of E is general recursive; in a similar way we extend the notion of general recursiveness to operations on expressions and relations between expressions. [11] Instead of *"general recursive"* we shall say for short *"recursive"*. Loosely speaking, a set E of expressions is recursive if there is a mechanical method which permits us to decide in each particular case whether a given expression belongs to E.

The exact nature of the one-to-one correspondence between expressions and natural numbers is to a large extent irrelevant for our discussion. The only important assumption which is to be made at this point is that, under the established correspondence, certain sets of expressions and operations on expressions prove to be recursive. This applies in particular to the sets of all variables, all formulas, all sentences, and all logical axioms; also to the operation of concatenating expressions (used in forming compound terms and formulas from simpler ones) and to the operations of inference. [12]

Furthermore, for an axiomatic theory we assume that the set of all its non-logical axioms, and hence also the set of all its axioms, is recursive. In this connection we agree to call an arbitrary theory

---

proved to be a powerful instrument in metamathematical investigations; cf. also the remarks in [31], pp. 301 f. and 404.

[11] A discussion of general recursiveness can be found, e.g., in [9], vol. 2, pp. 392 ff., and [15], pp. 73 ff.

[12] Instead of treating variables and constants as "atomic" symbols, we can choose two distinct symbols (to which no independent function in the structure of the theory is assigned) and construct the variables and constants, and hence also all compound expressions, as finite concatenations (successions) of these two symbols. In this case the only assumption to be satisfied by the correspondence between expressions and natural numbers is that, under this correspondence, the operation of concatenating two expressions should be recursive. In fact, if this condition is satisfied, it is easy to construct the variables and constants in such a way that all the sets of expressions and operations on expressions mentioned above in the text actually prove to be recursive. In particular this applies to the sets of logical axioms and the operations of inference which have been discussed in the literature.

T *axiomatizable* if there is a recursive set A of valid sentences of T such that every valid sentence of T is derivable from the set A; if the set A is assumed to be finite, the theory T is called *finitely axiomatizable*. (As is well known, a finite set is always recursive.) Thus every axiomatically built theory is axiomatizable in the sense just defined, and every axiomatizable theory can be represented as axiomatically built; a similar relation holds between finitely axiomatizable theories and those axiomatic theories in which the set of non-logical axioms is finite.

A theory T is called *decidable* if the set of all its valid sentences is recursive, and otherwise *undecidable*. This agrees with the intuitive explanation given in the introduction, according to which a theory T is decidable if there exists a decision procedure for T. In our further discussion we shall usually refer to this intuitive explanation rather than to the precise definition of a decidable theory in terms of recursive sets; hence our arguments will have an intuitive and informal character. Obviously, every decidable theory is axiomatizable; the converse, however, in general does not hold. A theory T is called *essentially undecidable* if not only T itself is undecidable, but the same holds for every consistent extension of T which has the same constants as T. Thus all theories fall into three mutually exclusive classes: decidable theories, essentially undecidable theories, and theories which are undecidable without being essentially undecidable. Instances of all three classes will be given below, in the remarks following Theorem 6.

In the next few theorems we formulate some elementary properties of undecidable and essentially undecidable theories. The first two of these theorems have some interest in themselves, but play no essential role in our further discussion.

THEOREM 1. *For a complete theory* T *the following three conditions are equivalent*: (i) T *is undecidable*, (ii) T *is essentially undecidable*, *and* (iii) T *is not axiomatizable*.

PROOF: It is known from the literature that, for every complete theory T, (i) implies (iii). [13] The remaining parts of the theorem

---

[13] This is a simple consequence of [13], p. 56, Theorem V; see also [10].

obviously follow from the definitions of the notions involved.

THEOREM 2. *For a theory T to be essentially undecidable it is necessary and sufficient that T be consistent and that no consistent and complete extension of T which has the same constants as T be axiomatizable.*

PROOF: The necessity of the condition obviously follows from Theorem 1 and the definitions of the notions involved. The statement that the conditon in question is sufficient is clearly equivalent to the following

LEMMA. *Every consistent and decidable theory T' has a consistent, complete, decidable (and hence also axiomatizable) extension T" which has the same constants as T'.*

To prove this lemma we start with an argument used in the proof of the well-known theorem (due to Lindenbaum) by which every consistent theory T' has a consistent and complete extension T"; cf. [32], Theorem I. 56.

In fact, we arrange all the sentences of T' into an infinite sequence $\langle \Phi_0, \ldots, \Phi_n, \ldots \rangle$. We then define recursively a sequence of natural numbers $\langle k_0, \ldots, k_n, \ldots \rangle$ by assuming that k_n is the smallest natural number p satisfying the conditions: (i) $k_i < p$ for every natural number $i < n$, and (ii) the sentence $\sim \Phi_p$ is not derivable from the set of all valid sentences supplemented by all the sentences Φ_{k_i} with $i < n$. Such a number p must exist. In fact, since there are infinitely many logically valid sentences in T', we can find a number p which satisfies (i) and for which Φ_p is logically valid; as is easily seen, (ii) will then be satisfied as well.

Let A be the set of all the sentences $\Phi_{k_0}, \ldots, \Phi_{k_n}, \ldots$. We easily show (with the help of Deduction Theorem I) that (i) every sentence which is derivable in T' from A belongs to A, (ii) every sentence which is valid in T' belongs to A, and (iii) of any two given sentences Ψ and $\sim \Psi$ in T' one and only one belongs to A. We can now define the theory T" by stipulating that the set of all its valid sentences coincides with A. Clearly T" is a complete and consistent extension of T' and has the same constants as T'.

Since the set of all sentences of T' is recursive, the sequence

$\langle\, \Phi_0,\, \ldots,\, \Phi_n,\, \ldots\, \rangle$ can also be assumed to be recursive. From the definition of A we conclude (again with the help of Deduction Theorem I) that a sentence Φ_p belongs to A if and only if there is a finite sequence of natural numbers $\langle\, k_0,\, \ldots,\, k_n\, \rangle$ satisfying the conditions:

(1) $k_0 < \ldots < k_n = p$;

(2) the sentence $\sim(\Phi_{k_0} \wedge \ldots \wedge \Phi_{k_n})$ is not valid in T';

(3) if $j < k_0$, then $\sim \Phi_j$ is valid in T';

(4) if $0 \leqslant i < n$ and $k_i \leqslant j < k_{i+1}$, then $\sim(\Phi_{k_0}\wedge \ldots \wedge \Phi_{k_i} \wedge \Phi_j)$ is valid in T'.

Hence, the theory T' being decidable, we can always decide in a finite number of steps whether or not a given sentence Φ_p belongs to A; in other words, the set A is recursive. Consequently, the theory T'' is decidable and *a fortiori* axiomatizable. This completes the proof of our lemma as well as of Theorem 2.

THEOREM 3. *Let* T_1 *and* T_2 *be two theories such that* T_1 *is a consistent extension of* T_2. *If* T_2 *is essentially undecidable, then* T_1 *is also essentially undecidable.*

PROOF: Let T_3 be the theory determined by the condition: Φ is a valid sentence of T_3 if and only if Φ is a sentence of T_2 and is valid in T_1. Clearly, T_3 is an extension of T_2 and has the same constants as T_2; moreover, T_3 is consistent since T_1, which is an extension of T_3, is consistent. Hence T_3 is undecidable. A decision procedure for T_1 would automatically yield a decision procedure for T_3. Consequently, T_1 is undecidable. The same argument obviously applies to every consistent extension of T_1, and therefore T_1 is essentially undecidable.

THEOREM 4. *Let* T_1 *and* T_2 *be two theories such that* T_2 *is an inessential extension of* T_1. *Then* T_1 *is undecidable, or essentially undecidable, if and only if* T_2 *is respectively undecidable, or essentially undecidable.*

PROOF: T_2 is obtained from T_1 by adding finitely or infinitely many new individual constants $c_0, c_1, \ldots, c_m, \ldots$. Given any sentence Φ of T_2, let $c_{m_1}, \ldots, c_{m_n}$ be all the new constants occurring in Φ. We choose n variables $x_1, \ldots, x_n$ which do not occur in Φ, and we replace each of the constants c_{mk}, wherever it occurs in Φ, by the corresponding variable x_k. If Φ' is the formula thus obtained, then

$$\wedge x_1 \ldots \wedge x_n \Phi'$$

is obviously a sentence of T_1; and from the definition of an in-essential extension we easily conclude that this sentence is valid in T_1 if and only if the sentence Φ is valid in T_2. Hence every decision procedure for T_1 leads to a decision procedure for T_2. On the other hand, if Ψ is an arbitrary sentence of T_1, we easily see that Ψ is valid in T_2 if and only if it is valid in T_1. Thus every decision procedure for T_2 automatically yields a decision procedure for T_1. Consequently, T_1 is undecidable if and only if T_2 is un-decidable. To obtain the analogous conclusion for essential unde-cidability, we notice that for every extension T_1' of T_1 there is an extension T_2' of T_2 such that T_2' is an inessential extension of T_1'. Hence we easily conclude that, if T_2 is essentially undecidable, T_1 is also essentially undecidable; the implication in the opposite direction follows from Theorem 3.

THEOREM 5. *Let* T_1 *and* T_2 *be two theories, with the same con-stants, such that* T_2 *is a finite extension of* T_1. *If* T_2 *is undecidable, then* T_1 *is also undecidable.*

PROOF: By the definition of a finite extension, there is a finite sequence of sentences $\Phi_1, \ldots, \Phi_n$ of T_2 such that a sentence is valid in T_2 if and only if it is derivable from the set of all valid sentences of T_1 supplemented by $\Phi_1, \ldots, \Phi_n$. Hence, by Deduction Theorem I, the problem whether a given sentence Ψ of T_2 is valid in T_2 is equivalent to the problem whether the sentence

$$(\Phi_1 \wedge \ldots \wedge \Phi_n) \to \Psi$$

is valid in T_1. Thus every decision procedure for T_1 leads to a decision procedure for T_2; and therefore, since T_2 is undecidable, T_1 cannot be decidable. [14]

---

[14] Essentially the same argument was applied to two special theories T_1 and T_2 in [2].

THEOREM 6. *Let* T_1 *and* T_2 *be two compatible theories such that every constant of* T_2 *is also a constant of* T_1. *If* T_2 *is essentially undecidable and finitely axiomatizable, then* T_1 *is undecidable, and so is every subtheory of* T_1 *which has the same constants as* T_1.

PROOF: Let T be the union of T_1 and T_2. By Theorem 3, T—as a consistent extension of T_2—is (essentially) undecidable. T and T_1 have obviously the same constants and, since T_2 is finitely axiomatizable, T is a finite extension of T_1. Therefore, by Theorem 5, T_1 is undecidable. If T_1' is a subtheory of T_1 which has the same constants as T_1, then T_1' and T_2 are clearly compatible and every constant of T_2 is a constant of T_1'; hence, by the same argument as before, T_1' is undecidable.

A theory T may be called *hereditarily undecidable* if, not only T itself, but also every subtheory of T which has the same constants as T is undecidable. Thus, under the hypothesis of Theorem 6, the theory T_1 proves to be hereditarily undecidable.

Various problems naturally arise which concern possible improvements of the results stated in Theorems 3–6. Thus, e.g., we can ask whether it is possible to remove the word *"essentially"* from Theorems 3 and 6, or to weaken the hypotheses of Theorem 5 by taking T_2 to be an arbitrary extension of T_1, or to omit in Theorem 6 the assumption that T_2 is finitely axiomatizable. The answer to these questions proves to be negative.

In fact, let T_1 be an arbitrary theory which has a binary predicate **P** among its non-logical constants, and in which the sentence

$$\wedge x \wedge y\,(x = y)$$

is accepted as the only non-logical axiom. (This sentence clearly expresses the fact that the universe consists of only one element; although non-logical constants do not occur in the sentence, it is not a logical axiom since it is not satisfied in every realization.) Let T_2 be the theory which has the same constants as T_1 but which has no non-logical axioms; every valid sentence of T_2 is logically valid. Thus T_2 is simply a system of the first-order predicate logic, with the constants restricted to those occurring in T_1. T_1 is a consistent extension of T_2, and hence the two theories are compatible. Clearly, T_2 is finitely axiomatizable; also it is known to be undecidable, while T_1 can easily be shown to be decidable. [15] Hence we see that Theorems 3 and 6 are no longer valid if we remove from them the word *"essentially"*.

On the other hand, an example of a decidable and finitely axiomatizable theory T_1 is known which has non-denumerably many consistent and complete extensions, all with the same constants as T_1. [16] Since, as is known, there are only denumerably many recursive sets of sentences in any given theory, we conclude that T_1 must have a consistent, complete, and undecidable extension T_2. T_1 and T_2 are of course compatible; T_2 is essentially undecidable (cf. Theorem 1), while T_1, as we have stated above, is decidable. This example shows that we cannot extend Theorem 5 to the case when T_2 is an arbitrary extension of T_1, and we cannot omit in Theorem 6 the assumption that T_2 is finitely axiomatizable.

From the point of view of applications it would be important to know whether Theorem 6 can be improved by assuming that T_2 is an arbitrary axiomatizable theory (which may not be finitely axiomatizable). We could answer this question affirmatively if we knew that every essentially undecidable theory which is axiomatizable has an essentially undecidable subtheory which is finitely axiomatizable. However, both problems are open. The example used above to show that the condition of finite axiomatizability in Theorem 6 cannot be simply omitted is not applicable to the new problems; for the essentially undecidable theory T_2 indicated in this example is complete and hence, by Theorem 1, is not axiomatizable.

---

[15] The decidability of T_1 can be shown by means of the well-known method of eliminating quantifiers which is discussed and used, e.g., in [30]; the application of this method to the decision problem for T_1 is extremely simple. The undecidability of an arbitrary system of predicate logic which, like T_2, has at least one binary predicate among its non-logical constants follows directly from the results in [2] and [12].

[16] The results in [26] and [27] show that the elementary theory of Abelian groups can be taken as T_1.

I.4. **Interpretability and weak interpretability.** To widen the range of application of Theorems 3–6 we introduce the notion of interpretability.

We want first to explain by means of examples what we understand by a *possible definition* of a given constant in a theory T to which this constant does not belong. Let the constant in question be a binary predicate, e.g., the less-than symbol $<$. A possible definition of $<$ in T is any sentence of the form

(i) $$\wedge x \wedge y(x < y \leftrightarrow \Phi)$$

where Φ is a formula of T (x and y being any two different variables such that no variable different from both of them occurs free in Φ). Clearly, (i) is not a formula in T, but it is a formula, and in fact a sentence, in every extension of T which contains $<$ as a constant. If the constant discussed is a binary operation symbol, e.g., the plus symbol $+$, then by a possible definition of this constant in T we understand a sentence of the form

(ii) $$\wedge x \wedge y \wedge z(x = y + z \leftrightarrow \Phi)$$

where Φ is again a formula of T. In this case, however, Φ is not quite arbitrary, for we assume that the following sentence is valid in T:

(iii) $$\wedge y \wedge z \vee x [\Phi \wedge \wedge u(\Phi' \rightarrow x = u)]$$

where u is an arbitrary variable (different from x, y, and z) which does not occur in Φ, and Φ' is the formula obtained from Φ by replacing x everywhere by u. Intuitively speaking, sentence (iii) expresses the fact that, in every possible realization of T, for any two elements represented by y and z there is just one element represented by x such that the three elements satisfy the formula Φ. We proceed analogously with predicates and operation symbols of arbitrary ranks; individual constants are treated in this context as operation symbols of rank 0.

Let now T_1 and T_2 be any two theories. First assume that T_1 and T_2 have no non-logical constants in common. In this case we say that T_2 is *interpretable* in T_1 if we can extend T_1, by including in the set of valid sentences some possible definitions of the non-logical constants of T_2, in such a way that the resulting extension of T_1 turns out to be an extension of T_2 as well. Speaking precisely, T_2 is interpretable in T_1 if and only if there is a theory T and a set D satisfying the following conditions: (α) T is a common extension of T_1 and T_2, and every constant of T is a constant of T_1 or T_2; (β) D is a recursive set of sentences which are valid in T and which are possible definitions in T_1 of non-logical constants of T_2; (γ) each non-logical constant of T_2 occurs in just one sentence of D; (δ) every valid sentence of T is derivable (in T) from a set of sentences each of which is valid in T_1 or belongs to D. In case T_2 has only finitely many non-logical constants, the term "recursive" in condition (β) can clearly be omitted. In the general case, when T_1 and T_2 may have some common non-logical constants, we first replace the non-logical constants in T_2 by new constants not occurring in T_1 (different symbols by different symbols), without changing the structure of T_2 in any other respect; if the resulting theory T_2' proves to be interpretable in T_1, we say that T_2 is also interpretable in T_1. It is easily seen that, in order to prove the interpretability of T_2 in T_1 in the general case, it is sufficient, though not necessary, to construct a theory T and a set D satisfying conditions (α)–(δ), with the difference that D is assumed to contain possible definitions of only those constants of T_2 which do not occur in T_1.

A theory T_2 is said to be *weakly interpretable* in T_1 if T_2 is interpretable in some consistent extension of T_1 which has the same constants as T_1. In order to prove the weak interpretability of T_2 in T_1 it suffices to show that T_1 and T_2 have a common consistent extension T satisfying the following condition: (ε) there is a recursive set D of valid sentences of T such that, for every non-logical constant C of T_2 which does not occur in T_1, some possible definition of C in T_1 belongs to D. In case T_2 has only finitely many non-logical constants, condition (ε) can clearly be replaced by a simpler one: (ζ) for every non-logical constant C of T_2 which does not occur in T_1, some possible definition of C in T_1 is valid in T. Moreover, in both (ε) and (ζ) the notion of a possible definition as applied to operation symbols and individual constants can be understood in a wider sense than the original one. If, e.g., the operation symbol $+$ does not occur in T_1, then as a possible definition of $+$ in T_1 we can regard any sentence of the form

$$\wedge x \wedge y \wedge z (x = y + z \leftrightarrow \varPhi)$$

where $\varPhi$ is a formula in T_1, independent of whether a sentence expressing the condition of existence and unicity for $\varPhi$ is valid in T_1. [17]

From the definition of interpretability we easily conclude that, if T_2 is interpretable in T_1, then every subtheory of T_2 is interpretable in every extension of T_1; also, if T_2 is interpretable in T_1 and T_3 is interpretable in T_2, then T_3 is interpretable in T_1. Analogous conclusions for weak interpretability fail; we can only show that, if T_2 is weakly interpretable in T_1, the same applies to every subtheory of T_2.

THEOREM 7. *Let T_1 and T_2 be two theories such that T_1 is consistent and T_2 is interpretable in T_1 or in some inessential extension of T_1. We then have:*

(i) *if T_2 is essentially undecidable, the same applies to T_1;*

(ii) *if T_2 has a finitely axiomatizable subtheory which is essentially undecidable, then so has T_1.*

PROOF: Consider first the case when T_2 is interpretable in T_1. Without loss of generality we can clearly assume that T_1 and T_2 have no common non-logical constants. Let T be a theory and D a set which satisfy conditions (α)–(δ) listed in the definition of interpretability. In view of (β) and (γ), using the sentences of D,

---

[17] The notions of interpretability and weak interpretability just defined can be generalized in several directions. In particular, under the definitions given in the text, the interpretation is restricted to non-logical constants; for various purposes it is desirable to have the definitions extended in such a way that the logical constants (especially the quantifiers and the identity symbol) could also be interpreted. It should be mentioned that we use here the term "weakly interpretable" in essentially the same sense in which the term "consistently interpretable" was used in earlier publications, e.g., in [34].

we transform every sentence Ψ of T into a well determined sentence Ψ^* of T_1 by eliminating from Ψ the non-logical constants of T_2.[18] We notice that $(\sim\Psi)^*$ coincides with $\sim(\Psi^*)$ and that the correlation between Ψ and Ψ^* is recursive. With the help of (α), (β), and (δ) we show that Ψ^* is valid in T_1 if and only if Ψ is valid in T. Hence, T_1 being consistent, T is also consistent. Thus, by (α), T is a consistent extension of the essentially undecidable theory T_2, and therefore T is undecidable (cf. Theorem 3). Furthermore, every decision procedure for T_1 automatically yields a decision procedure for T. Hence T_1 is undecidable. Clearly, T_2 being interpretable in T_1, it is also interpretable in every extension of T_1. Consequently, by the same argument as above, every consistent extension of T_1 is undecidable, and T_1 is essentially undecidable.

Assume now that T_2 has a finitely axiomatizable and essentially undecidable subtheory T_2'. A sentence which is derivable from a set X of sentences is always derivable from a finite subset of X, and conversely; this is clearly seen, e.g., from Deduction Theorem II. Hence, with the help of (α), (δ), and the definition of finite axiomatizability, we easily conclude that there is a finite set A of valid sentences of T_1 such that every valid sentence of T_2' is derivable (in T) from the union of A and D. Let T_1' be the theory in which the set of valid sentences consists of those and only those sentences which are derivable in T_1 from A. Clearly, T_1' is a finitely axiomatizable subtheory of T_1. We also see that T_2' is interpretable in T_1', and hence, by what we have already proved, T_1' is essentially undecidable.

With the help of Theorem 4 the conclusions obtained can easily be extended to the case when T_2 is interpretable, not in T_1, but in some inessential extension of T_1.

THEOREM 8.   *Let T_1 and T_2 be two theories such that T_2 is weakly interpretable in T_1 or in some inessential extension of T_1. If T_2 is essentially undecidable and finitely axiomatizable, then*:

---

[18] The elimination procedure is obvious so far as predicates are involved. It is but slightly more complicated in case the constants to be eliminated are operation symbols or individual constants; cf. an analogous argument in [9], vol. 2, pp. 144 ff.

(i) T_1 *is undecidable and every subtheory of* T_1 *which has the same constants as* T_1 *is undecidable*;

(ii) *there exists a finite extension of* T_1 *which has the same constants as* T_1 *and is essentially undecidable.*

PROOF: If T_2 is weakly interpretable in T_1, then, by definition, it is interpretable in some consistent extension T of T_1 which has the same constants as T_1. Consequently, by Theorem 7 (ii), there is a finitely axiomatizable subtheory T' of T which is essentially undecidable. T_1 and T' are clearly compatible, and every constant of T' is also a constant of T_1. Hence conclusion (i) follows immediately by Theorem 6. To obtain (ii) we consider the union T'' of T_1 and T'; since T' is finitely axiomatizable, T'' is a finite extension of T_1 and, by Theorem 3, T'' is essentially undecidable. With the help of Theorem 4, the conclusions easily extend to the case when T_2 is weakly interpretable in some inessential extension of T_1.

If a theory T_1 is an extension of a theory T_2, then T_2 is clearly interpretable in T_1; if T_1 and T_2 are compatible and every constant of T_2 is also a constant of T_1, then T_2 is weakly interpretable in T_1. Hence Theorems 7 and 8 are generalizations of Theorems 3 and 6, respectively. Also, remembering the remarks which follow Theorem 6, we see that Theorems 7 and 8 cannot be extended to arbitrary undecidable theories and that the assumption of finite axiomatizability in Theorem 8 cannot be omitted. Finally, we can show by means of examples that in general, under the hypothesis of Theorem 8, the theory T_1 is not essentially undecidable.

I.5. Relativization of quantifiers. The discussion in this section will further contribute to widening the applicability of our methods and results.

Given a theory T and a unary predicate **P**, we construct a new theory $T^{(P)}$ which will be referred to as the theory obtained by the *relativization of quantifiers in* T *to* **P** or, for short, by *relativizing* T *to* **P**. The set of all constants of $T^{(P)}$ consists of all constants of T and of the predicate **P**. If in a formula Φ of T we replace every subformula of the form $\wedge x \Psi$, or $\vee x \Psi$, by the expression $\wedge x(\mathbf{P}x \to \Psi)$, or $\forall x(\mathbf{P}x \land \Psi)$, respectively, then the resulting formula $\Phi^{(\mathbf{P})}$ is said to be obtained by *relativizing* Φ *to* $\mathbf{P}$. This definition of $\Phi^{(\mathbf{P})}$ could be formulated more precisely with the help of a recursive procedure, starting with atomic formulas and passing through all the operations by means of which compound formulas are constructed from simpler ones. We now define validity in $\mathsf{T}^{(\mathbf{P})}$ by stipulating that a sentence is valid in $\mathsf{T}^{(\mathbf{P})}$ if and only if it is derivable from the set of all sentences $\Phi^{(\mathbf{P})}$ correlated with sentences Φ which are valid in T. It is possible to give an equivalent definition of validity in $\mathsf{T}^{(\mathbf{P})}$ referring to models of T and not involving the construction of $\Phi^{(\mathbf{P})}$.

THEOREM 9. *Let* T *be any theory and* $\mathbf{P}$ *a unary predicate which is not a constant of* T. *We then have*:

(i) $\mathsf{T}^{(\mathbf{P})}$ *is axiomatizable if and only if* T *is axiomatizable*;

(ii) *under the assumption that only finitely many individual constants and operation symbols occur in* T, $\mathsf{T}^{(\mathbf{P})}$ *is finitely axiomatizable if and only if* T *is finitely axiomatizable*.

PROOF: Assume that T is axiomatizable. Thus there is a recursive set A of valid sentences of T such that every sentence which is valid in T is derivable from A. Let B be the set consisting of the following sentences: (I) all the sentences $\Phi^{(\mathbf{P})}$ correlated with sentences Φ of A; (II) the sentence $\forall x\, \mathbf{P}x$; (III) all the sentences of the forms

$$\mathbf{P}\mathbf{c}, \quad \wedge x[\mathbf{P}x \to \mathbf{P}(\mathbf{S}x)], \quad \wedge x \wedge y[(\mathbf{P}x \land \mathbf{P}y) \to \mathbf{P}(x + y)],$$

etc., where $\mathbf{c}$ is an individual constant, $\mathbf{S}$ a unary operation symbol, $+$ a binary operation symbol, etc., all occurring in T. (Sentence (II) may be omitted if at least one individual constant occurs in T; on the other hand, no sentences (III) are included in B if all non-logical constants of T are predicates.)

Since the correlation between Φ and $\Phi^{(\mathbf{P})}$ is recursive, the set B is also recursive. All sentences (I) are obviously valid in $\mathsf{T}^{(\mathbf{P})}$. The same applies to sentence (II). For let Φ be the sentence $\forall x(x = x)$. Φ is logically valid and hence valid in T. Therefore $\Phi^{(\mathbf{P})}$ is valid in $\mathsf{T}^{(\mathbf{P})}$; and $\Phi^{(\mathbf{P})}$ is logically equivalent to (II), in the sense that the sentence

$$\Phi^{(\mathbf{P})} \leftrightarrow \forall x\, \mathbf{P}x$$

is logically valid. Similarly we can show that all sentences (III) are valid in $\mathsf{T}^{(P)}$. For instance, let $+$ be a binary operation symbol occurring in T, and let Φ be the sentence $\wedge x \wedge y \vee z (z = x + y)$. Φ is logically valid and hence $\Phi^{(P)}$ is valid in $\mathsf{T}^{(P)}$; on the other hand $\Phi^{(P)}$ is logically equivalent to $\wedge x \wedge y [(\boldsymbol{P}x \wedge \boldsymbol{P}y) \to \boldsymbol{P}(x + y)]$. Thus every sentence of B is valid in $\mathsf{T}^{(P)}$.

We now want to prove that every sentence which is valid in $\mathsf{T}^{(P)}$ is derivable from B. This clearly amounts to showing that, if Φ is derivable from A, then $\Phi^{(P)}$ is derivable from B. We assume that the set of logical axioms has been selected in such a way that the operation of detachment suffices as the only operation of inference (cf. I. 2). By considering any such set of logical axioms known from the literature, we easily show that, if Φ is a logical axiom, then $\Phi^{(P)}$ is derivable from B; in fact, $\Phi^{(P)}$ is then derivable from sentences (II) and (III) alone. If Φ is in A, then $\Phi^{(P)}$ is obviously derivable from B. If Φ and Ψ are two sentences such that $\Phi^{(P)}$ and $(\Phi \to \Psi)^{(P)}$ are derivable from B, then $\Psi^{(P)}$ is derivable from B since $(\Phi \to \Psi)^{(P)}$ coincides with $\Phi^{(P)} \to \Psi^{(P)}$; thus the set of all sentences Φ such that $\Phi^{(P)}$ is derivable from B is closed under the operation of detachment. Hence this set contains all sentences derivable from A.

We have constructed a recursive set B of valid sentences of $\mathsf{T}^{(P)}$ such that every sentence which is valid in $\mathsf{T}^{(P)}$ is derivable from B. Consequently, $\mathsf{T}^{(P)}$ is axiomatizable.

Assume now, conversely, that $\mathsf{T}^{(P)}$ is axiomatizable. Thus there is a recursive set B' of valid sentences of $\mathsf{T}^{(P)}$ such that every sentence valid in $\mathsf{T}^{(P)}$ is derivable from it. With every sentence Φ of $\mathsf{T}^{(P)}$ we correlate a sentence Φ^* of T obtained by replacing in Φ every subformula of the form $\boldsymbol{P}\alpha$ (where α is an arbitrary term) by $\alpha = \alpha$. Let A' be the set of all sentences Φ^* corresponding to sentences Φ of B'.

Clearly the set A' is recursive. If Φ is a sentence of B', then Φ is valid in $\mathsf{T}^{(P)}$ and hence derivable from the set of all sentences $\Psi^{(P)}$ where Ψ is valid in T. Consequently, by Deduction Theorem II, there is a logically valid sentence of the form

$$(\Psi_1^{(P)} \wedge \ldots \wedge \Psi_n^{(P)}) \to \Phi$$

where $\Psi_1, \ldots, \Psi_n$ are valid in T. It is known that, whenever a sentence Θ is logically valid, then the sentence Θ^* is also logically valid. (This can be proved, e.g., by showing that, if Θ holds in every possible realization, the same applies to Θ^*.) Θ^* clearly coincides with

$$[(\Psi_1^{(P)})^* \wedge \ldots \wedge (\Psi_n^{(P)})^*] \to \Phi^*.$$

From the definitions of $\Psi^{(P)}$ and Ψ^* it is easily seen that every sentence Ψ of T is logically equivalent to $(\Psi^{(P)})^*$. Hence the sentences $(\Psi_1^{(P)})^*, \ldots, (\Psi_n^{(P)})^*$ are valid in T, and therefore Φ^* is valid in T. Consequently every sentence of A' is valid in T.

If Φ is any sentence valid in T, then $\Phi^{(P)}$ is valid in $T^{(P)}$ and therefore derivable from B'. By again applying Deduction Theorem II, we show that $(\Phi^{(P)})^*$, and hence also Φ, is derivable from A'. We conclude that T is axiomatizable.

We have thus established conclusion (i) of our theorem. Moreover we notice that, whenever the set A (in the argument outlined above) is finite, the set B is also finite, provided there are only finitely many individual constants and operation symbols occurring in T. Furthermore, if B' is finite, then A' is always finite. Consequently, conclusion (ii) also holds, and the proof is complete.

If T is an axiomatic theory, then, by Theorem 9, $T^{(P)}$ can also be represented as an axiomatic theory; from the proof of this theorem we learn how to construct an adequate axiom system for $T^{(P)}$ if the axiom system of T is known.

Supplementing Theorem 9 (ii), we can show that, whenever $T^{(P)}$ is finitely axiomatizable and consistent, T can contain only finitely many individual constants and operation symbols.

THEOREM 10. *Let* T *be any theory and* **P** *a unary predicate which is not a constant of* T. *Then* $T^{(P)}$ *is essentially undecidable if and only if* T *is essentially undecidable.*

PROOF: We want first to show that

(1) $T^{(P)}$ is interpretable in T.

For this purpose we construct an extension T^* of T by adding **P** to the set of constants of T and by stipulating that a sentence is valid in T^* if and only if it is derivable from the set of all valid sentences of T supplemented by the following sentence Θ:

$$\wedge x(\mathbf{P}x \leftrightarrow x = x).$$

Obviously, Θ is a possible definition of $\mathbf{P}$ in T. Φ being any sentence of T, it is easily seen that $\Phi^{(P)}$ is derivable from the set consisting of Φ and Θ. Hence we conclude that every sentence valid in $\mathsf{T}^{(P)}$ is also valid in T^*, i.e., that T^* is an extension of $\mathsf{T}^{(P)}$. Consequently, (1) holds.

If T is consistent, then, by (1), $\mathsf{T}^{(P)}$ must also be consistent. In fact, from the construction of T^* it follows that every sentence of T which is valid in T^* is also valid in T. Hence the consistency of T implies that of T^*: since T^* is an extension of $\mathsf{T}^{(P)}$, this in turn implies the consistency of $\mathsf{T}^{(P)}$. If T is inconsistent, two sentences Φ and $\sim\Phi$ are valid in T and therefore the sentences $\Phi^{(P)}$ and $(\sim\Phi^{(P)})$ are valid in $\mathsf{T}^{(P)}$; since $(\sim\Phi)^{(P)}$ coincides with $\sim\Phi^{(P)}$, $\mathsf{T}^{(P)}$ is also inconsistent. Thus,

(2) $\mathsf{T}^{(P)}$ is consistent if and only if T is consistent.

Now, if $\mathsf{T}^{(P)}$ is essentially undecidable, we conclude from (1) and (2), by applying Theorem 7 (i), that the same applies to T.

Assume, conversely, that T is essentially undecidable. Then T is consistent and hence, by (2), $\mathsf{T}^{(P)}$ is also consistent. Consider any consistent extension U of $\mathsf{T}^{(P)}$. Let S be the set of all sentences Φ of T such that the correlated sentence $\Phi^{(P)}$ is valid in U. If a sentence Ψ of T is derivable from the set S, then, by Deduction Theorem II, there is a logically valid sentence Ω of the form

$$(\Phi_1 \wedge \ldots \wedge \Phi_n) \rightarrow \Psi$$

where $\Phi_1, \ldots, \Phi_n$ are sentences of S. Clearly $\Omega^{(P)}$ coincides with the sentence

$$(\Phi_1^{(P)} \wedge \ldots \wedge \Phi_n^{(P)}) \rightarrow \Psi^{(P)}.$$

Since Ω is logically valid, it is valid in T; therefore $\Omega^{(P)}$ is valid in $\mathsf{T}^{(P)}$ and hence in U. By the definition of S, the sentences $\Phi_1^{(P)}, \ldots, \Phi_n^{(P)}$ are also valid in U. Hence $\Psi^{(P)}$ is valid in U, and Ψ is in S. Thus every sentence which is derivable from the set S belongs to this set. Consequently we can   construct a theory $\overline{\mathsf{T}}$, with the same constants as T, by stipulating that the set of all sentences which are valid in $\bar{T}$ coincides with S. Clearly $\bar{T}$ is an extension of T. The relativized theory $\bar{T}^{(P)}$ is a subtheory of U and therefore is consistent; hence, by (2), $\bar{T}$ is consistent. Thus $\bar{T}$ is a consistent extension of T and therefore is undecidable. Since the correlation between Φ and $\Phi^{(P)}$ is recursive, a decision procedure for U would automatically enable us to decide, for any given sentence Φ of T, whether $\Phi^{(P)}$ is valid in U, i.e., whether Φ is valid in $\bar{T}$. Hence U must be undecidable. Consequently, $T^{(P)}$ is essentially undecidable. Thus our theorem holds in both directions.

By analyzing the proof just outlined we easily see that, if a theory T is simply undecidable, and not necessarily essentially undecidable, then the relativized theory $T^{(P)}$ is also undecidable. The converse, however, by no means holds. In fact, it turns out that $T^{(P)}$ is as a rule undecidable. More precisely, it can be shown that $T^{(P)}$ is undecidable if only T is consistent and the first-order predicate logic, with the constants restricted to those of T, is undecidable. Thus, e.g., $T^{(P)}$ is undecidable whenever T is consistent and contains a binary predicate among its non-logical constants.

A theory T_2 is said to be *relatively interpretable*, or *relatively weakly interpretable*, in a theory T_1 if the correlated theory $T_2^{(P)}$ (obtained by relativizing T_2 to a predicate P which does not occur in T_2) is interpretable, or weakly interpretable, in T_1 in the usual sense. [19] It is possible to establish some general conditions under which a theory T_2 proves to be relatively interpretable in a theory T_1 and which do not involve the construction of the auxiliary theory $T_2^{(P)}$.

We are frequently confronted with a situation in which we can easily show that a certain theory T_2 known to be essentially un-

---

[19] As was pointed out in footnote 17, the definition of interpretability could be generalized so as to admit, in particular, the interpretability of quantifiers; the notion of interpretability thus generalized would comprehend relative interpretability as a particular case.

decidable (and which in addition may be finitely axiomatizable) is relatively interpretable, or relatively weakly interpretable, in a given theory T_1, while the proof that T_2 is interpretable, or weakly interpretable, in T_1 in the usual sense is either impossible or much more involved. In such a situation we cannot reach any conclusions regarding the undecidability of the theory T_1, or of its subtheories and extensions, by applying Theorems 7 and 8 directly, but we can still obtain the desired results by applying first Theorem 10 (and possibly also Theorem 9), and then Theorems 7 and 8. Thus Theorem 10 considerably enhances the applicability and efficiency of our methods.

I.6. **Examples and applications.** The theorems stated in this paper constitute theoretical foundations for a general method (described in I.1 as the "indirect method") with the help of which a negative solution of the decision problem for various special theories can be obtained. Theorem 8 is especially important from this point of view; for, given a finitely axiomatizable and essentially undecidable theory, this theorem enables us to establish the undecidability of various other theories which may be very distant in their mathematical content from the original theory. Of course, the applicability of this theorem, and hence the usefulness of the whole method, depends essentially on the fact that some finitely axiomatizable theories are now available which have been shown to be essentially undecidable without the help of our method and which admit interpretations in a great variety of other theories.

Several essentially undecidable theories have been discussed in the literature of the last 20 years. As the first example we want to mention the arithmetic of natural numbers, to which we shall refer as Theory N. This theory is regarded here as a theory with standard formalization. The set of its non-logical constants consists of two binary operation symbols, $+$ and $\cdot$, and possibly some other symbols, e.g., the individual constants 0 and 1, and the binary predicate $<$. A sentence of N is by definition valid if it is true under the assumption that the variables range over the set of all natural numbers and that the constants $+$, $\cdot$, etc. have their ordinary arithmetical meaning. Theory N is known to be complete and undecidable [20]; hence, in an obvious way, it is essentially undecidable (see Theorem 1). However, N is not axiomatizable (see again Theorem 1), and *a fortiori* not finitely axiomatizable; therefore Theorem 8 cannot be applied to it.

Among axiomatic subtheories of N the most important is Peano's arithmetic, which will be referred to as Theory P. It has the same constants as N. Its non-logical axioms are the well-known axioms of Peano, with recursive definitions of $+$ and $\cdot$ regarded as axioms; since, however, the induction principle in its whole generality cannot be formulated in the symbolism of P, it is replaced by an infinite set of axioms—in fact, by all the particular instances of this principle which can be formulated in P. As was stated in the introduction, Theory P has been shown to be essentially undecidable. P is of course axiomatizable, but it is not finitely axiomatizable, and thus again cannot be applied in derivations based upon Theorem 8. [21]

On the other hand, however, P is known to be interpretable or relatively interpretable (see I.4 and I.5) in various other theories and in particular in various axiomatic systems of set theory; hence, by Theorems 7 and 10, all these systems are essentially

---

[20] The completeness of N is a simple consequence of a precise definition of true sentences; see [31], p. 317. The undecidability of N follows immediately from the following two results:

(I) If V is the set of all natural numbers correlated with sentences which are valid in N, then V is not definable in N; i.e., there is no formula in N with one free variable that is satisfied by every number of V and by no other number.

(II) Every recursive set of natural numbers is definable in N.

For (I) see [31], pp. 370 ff. and 399, or [15], pp. 88 f. Under a suitably chosen definition of recursiveness, (II) is an immediate consequence of the definitions of the notions involved; see [15], pp. 73 ff.

[21] The proof that P is not finitely axiomatizable is given in [25]. It should be mentioned that the results in [3] imply the existence of undecidable (but not of essentially undecidable) subtheories of P which are finitely axiomatizable. Two examples of such theories result directly from the discussion in [2]; one of them is simply the predicate logic, with the constants restricted to those of P; neither of them, however, is essentially undecidable.

undecidable (assuming that they are consistent). Some of these systems are based upon finite sets of non-logical axioms; the best-known system of this kind is the one developed in Bernays [1]. [22] Thus we finally get examples of theories which are both essentially undecidable and finitely axiomatizable. Unfortunately, these are theories with a very rich mathematical content, they are not readily interpretable in other theories, and hence their usefulness for our purposes has been so far almost negligible.

Mostowski and the author found an example of a finitely axiomatizable and essentially undecidable subtheory Q of P; later this example was considerably simplified by R. M. Robinson. [23] Theory Q turned out to be very suitable for our method; its mathematical content is meager, and it can easily be interpreted or at least weakly interpreted in many different theories. Hence Theory Q has become a powerful instrument in the study of the decision problem; with its help it has proved to be possible to obtain a negative solution of this problem for a large variety of theories for which the problem had previously been open. For many applications the specific structure of Q is irrelevant; the only fact that matters is that P, and hence also N, has a finitely axiomatizable and essentially undecidable subtheory.

Thus, by applying Theorem 6, we arrive at once at the conclusion that every theory which is compatible with Q and has the same constants as Q is undecidable. Hence, in particular, every subtheory of N in which the set of constants includes $+$ and $\cdot$ is undecidable, and therefore N is hereditarily undecidable; this generalizes the results known from the literature which concern various special subtheories of N. Furthermore, N is known to be

---

[22] The system developed in [1] contains two kinds of variables with different ranges; hence it may seem doubtful whether this system can be presented as a theory with standard formalization in the sense of the present paper. However, a slight modification of the system suffices to eliminate these doubts.

[23] See [16] and [22]. A detailed discussion of Theory Q and its applications to various subtheories of the arithmetic of natural numbers and integers will be found in II.

relatively interpretable in the arithmetic of integers J, a theory constructed analogously to N, but in which the variables are assumed to range over the set of all integers. Hence, Q is also relatively interpretable in J. By applying Theorems 8–10 we conclude that every subtheory of J in which +, ·, and possibly some other symbols occur as non-logical constants is undecidable. As interesting examples of such subtheories we may mention the elementary theories of rings, commutative rings, and ordered rings. (When referring to the elementary theory of a class of mathematical systems we have in mind that part of the general theory of these systems which can be formalized within the first-order predicate logic.) Furthermore, by Theorem 7, Theory J must have some finitely axiomatizable subtheories which are essentially undecidable; the elementary theory of non-densely ordered rings has proved to be a subtheory of this kind.

By using essentially the same method R. M. Robinson has shown in [23] that the elementary theories of various special rings (i.e., various special complete and consistent extensions of the elementary theory of arbitrary rings) are undecidable. Julia Robinson in [20] has succeeded in proving that N—and hence also Q—is relatively interpretable in the arithmetic A of rational numbers, and that consequently all the subtheories of A (with + and · as non-logical constants) are undecidable; among these subtheories we find the elementary theories of fields and ordered fields. The undecidability of the elementary theories of groups, lattices, modular lattices, complemented modular lattices, and abstract projective geometries has been established by the author. [24] Grzegorczyk in [8] has obtained analogous results for distributive lattices, Brouwerian

---

[24] See [39] and [40]; a detailed proof of the undecidability of group theory will be given in III. It may be mentioned that in the abstracts [16], [39], and [40] various results concerning the undecidability of rings, groups, and lattices have been formulated for relativized theories in the sense of I.5, although the proofs clearly show that the results apply to non-relativized theories as well. Since, as was pointed out in I.5, the relativized theories are as a rule undecidable, the formulation of the results in these abstracts is defective (though formally correct); the results concerning non-relativized theories are essentially stronger.

algebras, and some related algebraic and geometric systems. [25]

Wanda Szmielew and the author have shown that Q is interpretable in a small axiomatic fragment S of set theory. [26] S has two non-logical constants: the unary predicate **E** denoting the property of being a set, and the binary predicate **ϵ** denoting the element relation. The set of non-logical axioms of S consists of three sentences which state respectively that (i) any two sets with the same elements are identical, (ii) there is a set with no elements, and (iii) for any two sets a and b there is a set c consisting of those and only those elements which are elements of a or are identical with b. As opposed to more comprehensive systems of set theory, S can easily be shown to be consistent. Hence, by Theorem 7, S is essentially undecidable; and, by Theorem 3, the same applies to every consistent theory in which the non-logical axioms of S are valid. This directly implies that all axiomatic systems of set theory, with **E** and **ϵ** as non-logical constants, which are known from the literature are essentially undecidable (assuming that they are consistent); for each of these systems is an extension of S. The results easily extend to systems of set theory in which **ϵ** occurs as the only non-logical constant; S is then replaced by a closely related system S′ the three axioms of which are obtained from those of S by eliminating the constant **E**. By Theorem 5, the undecidability of S′ immediately implies that of the first-order predicate logic with a binary predicate as the only non-logical constant—a result which was previously obtained by a different method. [27]

To conclude we should like to mention an application of essentially undecidable and finitely axiomatizable theories to a decision problem of a different character (so to speak, to a decision problem of the second degree). [28] We have in mind the problem of the

---

[25]   In particular, Grzegorczyk gives a simple proof of undecidability for the elementary theory of closure algebras—a result originally obtained by Jaśkowski by means of a more involved method.

[26]   This and some related results are stated (without proof) in [28].

[27]   See footnote 15.

[28]   Still another application of these theories is indicated in [29].

existence of a method which would permit us in each particular case to decide whether or not a given theory is decidable. A precise statement of this problem involves some difficulties if one considers theories of a quite arbitrary character. To avoid these difficulties, we can restrict ourselves to axiomatic theories with standard formalization, build upon finite systems of non-logical axioms; for such theories our problem can easily be formulated in terms of recursive sets of sentences. By means of the following simple argument we can now show that the solution of the problem is negative even for this restricted class of formalized theories. Let T be any essentially undecidable and finitely axiomatizable theory; e.g., we take Q or S for T. Given a sentence Φ of T, let $T(\Phi)$ be the theory obtained from T by adding Φ to the set of axioms; obviously, $T(\Phi)$ is still finitely axiomatizable. Since T is essentially undecidable, the following three conditions are clearly equivalent for every sentence Φ: (i) $T(\sim \Phi)$ is decidable, (ii) $T(\sim \Phi)$ is not consistent, and (iii) Φ is valid in T. Hence an affirmative solution of our problem would imply the existence of a decision procedure for T, which is of course impossible.

In addition to decision problems of the types discussed in this paper, many other decision problems are known from the literature. This applies in particular to what may be called *restricted decision problems* for various theories T, i.e., to problems of determining whether a set S of all valid sentences of T satisfying certain additional conditions is recursive. For instance, by taking for T the arithmetic N, and for S the set of all valid sentences of N of the form

$$\mathsf{V}x_1 \ldots \mathsf{V}x_n (\alpha = \beta),$$

where α and β are two arbitrary terms, we arrive at the famous tenth problem of Hilbert; as another example we may mention the well-known word problem for groups. [29] It would be interesting and important to know whether and how our general method can be adapted to the study of decision problems of this kind.

---

[29] Various restricted decision problems for predicate logic are discussed in [4]. For a formulation of the word problem see III, p. 86.

II

UNDECIDABILITY
AND ESSENTIAL UNDECIDABILITY
IN ARITHMETIC

BY

ANDRZEJ MOSTOWSKI, RAPHAEL M. ROBINSON,
AND ALFRED TARSKI

II

UNDECIDABILITY AND ESSENTIAL UNDECIDABILITY IN ARITHMETIC

II.1. A summary of the results; notation. In this paper we concern ourselves with the decision problem for the formalized arithmetic of natural numbers and its subtheories which contain the symbols of arithmetical addition and multiplication among their non-logical constants. Our main results can be formulated as follows:

(I)   *There is a finitely axiomatizable subtheory of the arithmetic of natural numbers which is essentially undecidable.*

(II)   *Every subtheory of the arithmetic of natural numbers is undecidable.*

These two results are subsequently extended to the arithmetic of arbitrary integers. In particular, we show that the elementary theory of non-densely ordered rings presents an example of a finitely axiomatizable subtheory of the arithmetic of integers which is essentially undecidable.

The result (I) is obtained by means of what was referred to in I.1 as the direct method in proofs of undecidability. From this, the remaining results are derived by means of the indirect method discussed in I.3–5 in a detailed way. [1]

---

[1] A finitely axiomatizable and essentially undecidable subtheory $\overline{Q}$ of the arithmetic of natural numbers was first constructed by Mostowski and Tarski (in 1939); they also indicated the applications of their result to the decision problem for arbitrary subtheories of the arithmetic of natural numbers and integers as well as for various theories of rings. This theory $\overline{Q}$ will not be directly involved in the present discussion; it is closely related to the theory of non-densely ordered rings which will be discussed in II.6; cf. [16], and also [20], pp. 110 f., where a description of an analogous theory of positive integers can be found. Dr Julia Robinson, who read the manuscript originally prepared by Mostowski and Tarski, suggested that the treat-

The general plan of the paper can be described as follows: In II.2 we first introduce the notion of definability in an arbitrary formalized theory T; more specifically, we explain under what condition a function on and to the natural numbers, or a set of natural numbers, is said to be definable in T. We then show that there is no consistent theory T in which a certain function D (the diagonal function) and a certain set V (the set of numbers correlated with valid sentences of T) are both definable. Hence, under some additional, though still very general, assumptions, we conclude that every consistent theory in which all recursive functions are definable is essentially undecidable. These two results can serve as a general theoretical basis for the direct method in proofs of undecidab lity.

In II.3 we describe the formalized arithmetic N of natural numbers and some of its axiomatic subtheories — P, Q, and R. P is Peano's arithmetic; Q is a subtheory of P based upon a system of seven simple axioms; R is a subtheory of Q with a very weak, though infinite, axiom system. In II.4 we show that all recursive

---

ment of general recursive functions can be simplified by applying a characterization of these functions from which the recursive scheme has been eliminated; her suggestion has been followed in the present version of the proof of Theorem 6 in II.4. R. M. Robinson has shown that Theory $\overline{Q}$ can be replaced by a weaker theory based upon a simpler axiom system, in fact, by Theory Q described in II.3 (and that the decision problem for the latter reduces to that for a still weaker theory R, which, however, is no longer finitely axiomatizable); cf. [22]. Also Theorem 11 in II.5 concerning the irreducible character of the axiom system of Q is due to him. Moreover, he has shown that the part of the original proof which in the present version corresponds to the proof of Theorem 1 in II.2 can be considerably simplified. By analyzing the simplification suggested by R. M. Robinson and by comparing it with his own argument in [31], pp. 370 ff., and with Mostowski's extension of this argument in [15], p. 88, Tarski has obtained the general formulations given in II.2 as Theorem 1 and Corollary 2; see [37]. Finally it may be mentioned that, while the essential undecidability of Q directly implies that of the original theory $\overline{Q}$, the implication in the opposite direction can also be rather easily established by means of the general interpretation method discussed in I; however, this result, due to Wanda Szmielew and Tarski (see [28]), will not be applied in the present article.

functions are definable in R and hence also in every extension of R, e.g., in Q. By combining the results of II.2 and II.4, we establish in II.5 the main results of this paper quoted above as (I) and (II). In particular, Q provides us with a simple example of a subtheory of the arithmetic of natural numbers which is both finitely axiomatizable and essentially undecidable. In addition, we show that the axiom system of Q has a certain irreducibility property: by omitting any one axiom of this system we arrive at a subtheory of Q which is no longer essentially undecidable. Finally, in II.6 the main results of II.5 are extended to systems of the arithmetic of natural numbers with different sets of constants, as well as to the formalized arithmetic of integers. Among subtheories of the latter we find elementary theories of various kinds of rings (arbitrary rings, commutative rings, ordered rings, etc.), which are discussed in a more detailed way.

As regards the notation, the formula $x,y, \ldots \in Z$, or $x,y, \ldots \notin Z$, expresses the fact that the elements $x,y, \ldots$ belong, or do not belong, to the set Z. The ordered n-tuple whose consecutive terms are $x_1, \ldots, x_n$ (n being any natural number) is denoted by $\langle x_1, \ldots, x_n \rangle$. $X \times Y$ (the Cartesian product of X and Y) is the set of all ordered couples $\langle x,y \rangle$ such that $x \in X$ and $y \in Y$. The set of all natural numbers is denoted by N, and that of all integers by I. When applied to natural numbers and integers, the symbols 0, 1, $<$, $\leq$, $+$, $\cdot$, $\ldots$ are used in their ordinary meaning; the symbol S denotes the successor function (operation), i.e., $Sn = n + 1$ for every n. However, the same symbols 0, $<$, $+$, S, $\ldots$ are sometimes used to denote certain elements, relations between elements, and operations on elements of an abstract set. When referring to functions we shall usually have in mind functions with one argument on the set N and to the set N.

All the theories discussed are regarded as theories with standard formalization. [2] When speaking of constants and axioms of a theory we shall always mean non-logical constants and non-logical axioms.

---

[2] A description of theories with standard formalization is given in I.2.

It has proved convenient and desirable to distinguish in this paper between mathematical symbols used in their proper sense and metamathematical denotations of these symbols when the latter are treated as components of a formalized theory. The symbols printed in bold type are always used as metamathematical designations; in particular, $=$, $+$, and $\cdot$ are respectively used to denote the symbols of identity, addition, and multiplication occurring in the theories discussed. Thus, e.g., $\alpha = \beta$ denotes the equation with the terms α and β; when writing

$$\Phi = (\alpha = \beta)$$

we state the fact that the expression Φ coincides with this equation. On the other hand, sentential connectives and quantifiers will never be used in our discussion in their proper meanings and hence there is no need for distinguishing between these symbols and their metamathematical denotations. For instance, the expression

$$\Phi \rightarrow \Psi$$

is always to be read "the implication with the hypothesis Φ and the conclusion Ψ", and never "if Φ, then Ψ". [3]

To simplify the metamathematical symbolism, we accept certain conventions clarified by the following examples:

(i)   Instead of

$$\sim (\alpha = \beta),$$

where α and β are arbitrary terms, we write

$$\alpha \neq \beta.$$

(ii)   Instead of

$$(\Phi_1 \vee \Phi_2) \vee \Phi_3, \text{ or } (\Phi_1 \wedge \Phi_2) \wedge \Phi_3,$$

where Φ_1, Φ_2, and Φ_3 are formulas, we write

$$\Phi_1 \vee \Phi_2 \vee \Phi_3, \text{ or } \Phi_1 \wedge \Phi_2 \wedge \Phi_3.$$

(iii)   Instead of

$$\Phi_1 \rightarrow (\Phi_2 \vee \Phi_3), \text{ or } (\Phi_1 \vee \Phi_2) \leftrightarrow (\Phi_3 \wedge \Phi_4),$$

---

[3] Cf. I, footnote 5.

we write

$$\Phi_1 \rightarrow \Phi_2 \vee \Phi_3, \text{ or } \Phi_1 \vee \Phi_2 \leftrightarrow \Phi_3 \wedge \Phi_4.$$

(iv) Let Φ be a formula with one free variable x, Ψ a formula with two free variables x and y, etc. Instead of

$$\wedge x \Phi, \ \wedge x \wedge y \Psi, \ \ldots$$

we shall simply write

$$\Phi, \Psi, \ldots$$

if it is clear from the context that the expressions involved are supposed to be sentences.

(v) Let u and v be two distinct variables which are regarded as fixed in the whole subsequent discussion. Given two expressions Φ and α, we denote by $\Phi(\alpha)$ the expression obtained from Φ by replacing all occurrences of u by α; similarly, given three expressions Φ, α, and β, we denote by $\Phi(\alpha,\beta)$ the expression obtained from Φ by replacing simultaneously all occurrences of u by α and all occurrences of v by β. The symbols

$$\Phi, \ \Phi(u), \ \Phi(u,v)$$

clearly denote the same expression; we shall use the three symbols interchangeably, depending on which of these symbols seems the most suggestive in a given situation.

In most cases in which an expression $\Phi(\alpha)$ is involved it is assumed, explicitly or implicitly, that Φ is a formula, α a term, and that neither u nor any variable occurring in α occurs bound in Φ; $\Phi(\alpha)$ is then called a *regular substitution* of Φ. Without this (or a somewhat less stringent) assumption some passages in our further discussion lose their validity; in fact, all these passages which are based upon the fact that

$$\wedge u \Phi(u) \rightarrow \Phi(\alpha)$$

is a logically valid sentence. (However, no such assumption is involved in the definition of function D in II.2.) Similar remarks apply to the expression $\Phi(\alpha,\beta)$.

It may be mentioned that, in order to avoid complications

connected with the notion of a regular substitution, we could accept a different definition of $\Phi(\alpha)$ and $\Phi(\alpha,\beta)$, which would also be suitable for our purposes. In fact, we could agree to use $\Phi(\alpha)$ as an abbreviation for the expression

$$\wedge u(u = \alpha \to \Phi)$$

in case u does not occur in α, and for the expression

$$\wedge w[w = \alpha \to \wedge u(u = w \to \Phi)]$$

in case u occurs in α while w is a variable which does not occur either in Φ or in α. We may assume that all the variables have been arranged in an infinite sequence, and choose as w the first variable in this sequence occurring neither in Φ nor in α. $\Phi(\alpha,\beta)$ is defined analogously. Under this alternative definition the sentence

$$\wedge u\Phi(u) \to \Phi(\alpha)$$

is logically valid for every formula Φ and every term α. [4]

II.2. **Definability in arbitrary theories.** Let $\top$ be any theory with standard formalization. [5] We assume that in this theory an infinite sequence of terms $\Delta_0, \Delta_1, \ldots, \Delta_n, \ldots$ containing no variables is available.

A subset P of N, i.e., a set P of natural numbers, is said to be *definable* in $\top$ (relatively to the sequence of terms $\Delta_0, \Delta_1, \ldots$) if there is a formula Φ of $\top$ such that $\Phi(\Delta_n)$ is a sentence valid in $\top$ whenever $n \in P$, and $\sim\Phi(\Delta_n)$ is a sentence valid in $\top$ whenever $n \in N$ and $n \notin P$. [6] Such a formula Φ is said to *define* P. From the fact that $\Phi(\Delta_n)$ is a sentence we conclude (looking back at the

---

[4] Compare [35].

[5] Actually the discussion in this section applies to arbitrary formalized theories whose logical basis includes at least the first-order predicate logic with identity. With small changes, the discussion could even be extended to theories in which no quantifiers occur.

[6] The notion of definability was first discussed in [36]. A more general notion, which comprehends both the notion of definability in the sense of [36] and that of general recursiveness as particular cases, was introduced in [15], pp. 72 ff.; it is still not as general as the notion discussed in this paper.

definition of $\Phi(x)$ in II. 1) that the variable u does not occur bound in Φ and that no variable different from u occurs free in Φ.

A function F is said to be *definable* in $\top$ if there is a formula Φ such that

(i) for any $n,p \in N$ with $Fn = p$, $\Phi(\varDelta_n, \varDelta_p)$ is a sentence valid in $\top$;

(ii) for any $n,p \in N$ with $Fn \neq p$, $\sim\Phi(\varDelta_n, \varDelta_p)$ is a sentence valid in $\top$;

(iii) for every $n \in N$,

$$\wedge u \wedge v[\Phi(\varDelta_n, u) \wedge \Phi(\varDelta_n, v) \rightarrow u = v]$$

is a sentence valid in $\top$.

As before, such a formula Φ is said to *define* F; the variables u and v do not occur bound in Φ, and no variable different from u and v occurs free in Φ. It is easily seen that conditions (i) and (iii) can be replaced by the following condition:

(i') for every $n \in N$, the sentence

$$\wedge v[\Phi(\varDelta_n, v) \leftrightarrow v = \varDelta_{Fn}]$$

is valid in $\top$.

Under the assumption that, for all $n,p \in N$ with $n \neq p$, the sentences $\varDelta_n \neq \varDelta_p$ are valid in $\top$, condition (ii) proves to be a consequence of (i) and (iii), and hence can be dispensed with. Even without this assumption, however, the omission of condition (ii) would not invalidate the proof of the essential results of this section, i.e., Theorem 1 and Corollary 2.

By modifying in an obvious way the definition of definable sets, we arrive at the notion of definable binary relations (sets of ordered couples) and, more generally, of definable n-ary relations. Since functions (with one argument) can be treated as special binary relations, we obtain in this way a new definition of definable functions. The new definition is weaker than the original one and differs from the latter by the absence of condition (iii). It suffices for the proof of Theorem 3 and Corollary 4 outlined below; however, condition (iii) will be essentially involved in the proof of Theorem 1 and Corollary 2. When using the term "definable functions" in the present article, we shall have in mind the original definition of this term.

On the other hand, the definability of sets reduces under certain conditions to the definability of functions. In fact, let P be any set of natural numbers, and let C_P be its characteristic function defined by the formulas:

$$C_P n = 0 \text{ if } n \in P,$$
$$C_P n = 1 \text{ if } n \in N \text{ and } n \notin P.$$

The definability of C_P (even in the weaker sense) implies that of P; in fact, if a formula Φ defines C_P, then the formula $\Phi(u, \Delta_0)$ is easily seen to define P. Under the assumption that, for all $n, p \in N$ with $n \neq p$, the sentences $\Delta_n \neq \Delta_p$ are valid in T, it turns out that, conversely, the definability of P implies that of C_P; for, if a formula Ψ defines P, then the formula

$$[\Psi(u) \wedge v = \Delta_0] \vee [\sim \Psi(u) \wedge v = \Delta_1]$$

proves to define C_P.

Consider now any fixed one-to-one correspondence between natural numbers and expressions of T. For the time being we make no other assumptions regarding the nature of this correspondence. Given a natural number n, the correlated expression of T will be denoted by E_n; conversely, given an expression Φ, we shall denote the correlated natural number by $Nr(\Phi)$.

Let D (the *diagonal function*) be the function defined by the formula

$$Dn = Nr(E_n(\Delta_n));$$

thus Dn is the number correlated with the expression which is obtained from E_n by replacing everywhere the variable u by the term Δ_n. Let V be the set of all natural numbers n such that E_n is a sentence valid in T. Using this notation we can establish the following result:

THEOREM 1. *If the theory T is consistent, then the function D and the set V are not both definable in T.*

PROOF: Assume, to the contrary, that both D and V are definable in T. Thus there are formulas Φ and Ψ satisfying, for every natural number n, the following conditions:

(1) the sentence

$$\wedge v[\Phi(\varDelta_n, v) \leftrightarrow v = \varDelta_{Dn}]$$

is valid in T;

(2)   if $n \in V$, then the sentence $\Psi(\varDelta_n)$ is valid in T;

(3)   if $n \notin V$, then the sentence $\sim \Psi(\varDelta_n)$ is valid in T.

The variable v does not occur free in Ψ. Without any loss of generally we can assume that v does not occur in Ψ at all and that, consequently, $\Psi(v)$ is a regular substitution of Ψ.

Let

$$m = Nr(\wedge v[\Phi(u, v) \to \sim \Psi(v)])$$

so that

$$E_m = \wedge v[\Phi(u, v) \to \sim \Psi(v)]$$

and consequently,

(4) $E_m(\varDelta_m) = \wedge v[\Phi(\varDelta_m, v) \to \sim \Psi(v)].$

If the sentence $E_m(\varDelta_m)$ is valid in T, then, by (1) (with $n = m$) and (4), the sentence $\sim \Psi(\varDelta_{Dm})$ is valid. If $E_m(\varDelta_m)$ is not valid, then $Nr(E_m(\varDelta_m)) \notin V$; and since, by the definition of the function D,

(5) $Dm = Nr(E_m(\varDelta_m)),$

we conclude by (3) that in this case $\sim \Psi(\varDelta_{Dm})$ is valid as well. Thus,

(6)   the sentence $\sim \Psi(\varDelta_{Dm})$ is valid in T.

By (1) and (6), the sentence

$$\wedge v[\Phi(\varDelta_m, v) \to \sim \Psi(v)]$$

is valid. Consequently, by (4) and (5), $Dm \in V$; and therefore, by (2),

(7)   the sentence $\Psi(\varDelta_{Dm})$ is valid in T.

(6) and (7) imply that the theory T is inconsistent, contrary to the hypothesis of the theorem. This completes the proof.

Theorem 1 and its proof represent a metamathematical reconstruction and generalization of arguments involved in various semantical antinomies and, in particular, in the antinomy of the liar. The idea of this reconstruction and the realization of its far-reaching implications is due to Gödel [7]. The present version of this reconstruction is distinguished by its generality and simplicity. It applies to arbitrary formalized theories, and not only to those in which a comprehensive fragment of the arithmetic of natural numbers can be developed; to a large extent it is independent of the way in which the notion of validity has been defined for a given theory, and in particular it does not involve the notion of a formal proof within this theory; it does not use the apparatus of recursive functions—although this apparatus will play a funda-mental role in applications of Theorem 1 to the decision problem.

From now on we assume that the correspondence between natural numbers and expressions of T satisfies the usual conditions con-cerning the recursiveness of certain functions and sets under this correspondence (cf. I.3). The most fundamental of these conditions can be expressed as follows: Let $\Phi^\frown\Psi$ denote the concatenation of two arbitrary expressions Φ and Ψ; then the function G on $N\times N$ to N defined by the formula

$$G(n,p) = Nr(E_n^\frown E_p)$$

is recursive. We also assume that the terms $\Delta_0, \Delta_1, \ldots$ form a recursive sequence, i.e., that the function H (on N to N) defined by the formula

$$Hn = Nr(\Delta_n)$$

is recursive. From these assumptions it follows, in particular, that the function D is also recursive. [7]

---

[7] Actually the recursiveness of D is the only assumption involved in our further discussion. A proof of this assumption can be obtained, e.g., from the discussion in [15], pp. 73 ff. It may be interesting to notice that, under the alternative definition of $\Phi(\alpha)$ mentioned at the end of II.1, the function D is defined by the formula

$$Dn = Nr[\wedge u(u = \Delta_n \to E_n)].$$

Using this formula and applying some elementary properties of recursive functions we can immediately derive the recursiveness of D from that of the functions G and H defined above in the text.

There are close connections between the notions of recursiveness and definability. In II.4 we shall see that many theories T are known such that all the functions and sets which are recursive are definable in T (and conversely). The importance of this fact for our discussion is seen from the following corollary which is an easy consequence of Theorem 1:

COROLLARY 2. *If* T *is a consistent theory in which all recursive functions are definable, then* T *is essentially undecidable.*

PROOF: As has been stated above, the function D is recursive and therefore, by the hypothesis of the corollary, it is definable in T. Hence, by Theorem 1, the set V is not definable in T. If the set V were recursive, its characteristic function C_V ($C_V n = 0$ for $n \in V$, $C_V n = 1$ for $n \notin V$) would also be recursive and hence definable in T; as was mentioned before, this would imply the definability of the set V itself. Thus, the set V is not recursive; in other words, the set of all sentences which are valid in T is not recursive, and the theory T is undecidable. Since the definability in T implies the definability in every extension of T (relatively to the same sequence of terms $\Delta_0, \Delta_1, \ldots$), our argument shows that every consistent extension of T is undecidable and that, consequently, T is essentially undecidable.

To conclude this section, we give two results, Theorem 3 and Corollary 4, concerning those theories in which every, or not every, definable function is recursive. These results follow in a simple way from the definitions of the notions involved and are less deep than Theorem 1 and Corollary 2; also they will find no essential applications in our further discussion. Nevertheless it will be interesting to compare the content of these results with that of Corollary 2.

THEOREM 3. *Every function which is definable in a consistent and axiomatizable theory* T *is recursive. The same applies to every set of natural numbers.*

PROOF: Using the intuitive notion of recursiveness, we argue as follows. Since T is axiomatizable, there is a recursive set A of sentences of T such that a sentence is valid in T if and only if it is derivable from A. Hence, as is well known, we can arrange all valid sentences of T in a recursive infinite sequence $\Psi_0, \Psi_1, \ldots$. (In other words, there is a recursive function the range of which is the set V.) Let now F be any function definable in T, and let Φ be a formula defining F. Then, for any given $n \in N$, the sentence $\Phi(\Delta_n, \Delta_{Fn})$ is valid in T, and therefore it must occur somewhere in the sequence $\Psi_0, \Psi_1, \ldots$. Hence, by checking the successive terms of this sequence, we must arrive in finitely many steps at a term Ψ_m which is of the form $\Phi(\Delta_n, \Delta_p)$ for some $p \in N$. If the numbers p and Fn were different, then the sentence $\sim \Phi(\Delta_n, \Delta_p)$ would be valid in T, and therefore T would be inconsistent; hence $p = Fn$. We thus have a procedure which enables us, for any given $n \in N$, to determine the function value Fn in finitely many steps; in other words, F is recursive. If, instead of a function, we consider a set of natural numbers, the proof remains practically unchanged.

The converse of Theorem 3 is by no means true. In fact, as is seen from Corollary 2, if T is any consistent theory which is decidable or, more generally, which is not essentially undecidable, then there are recursive functions which are not definable in T.

COROLLARY 4. *If* T *is a consistent theory such that there is a function (or set of natural numbers) which is definable in* T *but not recursive, then* T *is essentially undecidable.*

PROOF: By Theorem 3, the theory T is not axiomatizable and hence *a fortiori* undecidable (cf. I.3). For the same reason every consistent extension of T is undecidable, and hence T is essentially undecidable.

Corollaries 2 and 4 seem to provide us with two widely different methods of constructing essentially undecidable theories. Actually, however, we know no single case in which Corollary 4 would be of real help in a proof of essential undecidability. This is explained partly by the rather trivial character of Corollary 4 and partly by the fact that a theory T which satisfies the hypothesis of this corollary must have a much stronger property than essential undecidability; in fact, by analyzing the proof of Theorem 3, we conclude that, in every consistent extension of T, the set of all valid sentences not only is not recursive, but cannot even be recursively enumerable. [8]

II.3. Formalized arithmetic of natural numbers and its subtheories.
The formalized *arithmetic of natural numbers* in which we are interested here will be referred to as Theory N. The set of all constants of N is assumed to consist of four symbols: an individual constant **0**, a unary operation symbol **S**, and two binary operation symbols, **+** and **·**. The same assumption applies to all other theories discussed in this and the following two sections. Hence every possible realization of each of these theories is a system $\mathfrak{R} = \langle U, c, F, +, \cdot \rangle$ in which U is an arbitrary set, c is an element of U, F is a unary operation (function) on U to U, and $+$ and $\cdot$ are two binary operations on $U \times U$ to U (cf. I.2). To define the validity in N we consider a special realization $\mathfrak{R} = \langle N, 0, S, +, \cdot \rangle$ where all the symbols $N, 0, S, \ldots$ have their usual arithmetical meaning (cf. II.1); a sentence is said to be valid in N if and only if it holds in $\mathfrak{R}$.

We shall be interested in some axiomatic subtheories of N referred to as Theories P, Q, and R. The axiom system of Q consists of the following seven sentences:

Θ_1. $Sx = Sy \rightarrow x = y.$

Θ_2. $0 \neq Sy.$

Θ_3. $x \neq 0 \rightarrow \vee y(x = Sy).$

Θ_4. $x + 0 = x.$

Θ_5. $x + Sy = S(x + y).$

Θ_6. $x \cdot 0 = 0.$

Θ_7. $x \cdot Sy = (x \cdot y) + x.$

This axiom system is distinguished by the simplicity and the clear mathematical content of its axioms. Axiom Θ_1 expresses a fundamental property of **S**. Axioms Θ_2 and Θ_3 form together an

---

[8] For a definition of recursively enumerable sets see, e.g., [15], p. 86.

explicit definition of 0 in terms of S; they can be replaced by one equivalence

$$x = 0 \leftrightarrow \wedge y(x \neq Sy).$$

Axioms Θ_4 and Θ_5 constitute a recursive definition of $+$, and Axioms Θ_6 and Θ_7 a recursive definition of $\cdot$.

Theory P is known as *Peano's arithmetic*. Its axiom system consists of six individual sentences Π_1, Π_2, Π_4-Π_7, which respectively coincide with Θ_1, Θ_2, Θ_4-Θ_7, and moreover of an infinite system of sentences—in fact, of all sentences which are particular instances of the following axiom scheme of induction:

$\Pi_3.$ $\qquad\qquad \Phi(0) \wedge \wedge u[\Phi(u) \rightarrow \Phi(Su)] \rightarrow \wedge u\Phi(u)$

where Φ is an arbitrary formula in which u does not occur bound. [9]

To describe conveniently the axioms of R, and also for further use, we introduce the following notation.

Consider the infinite sequence of terms

$$0, \, S0, \, SS0, \, \ldots .$$

The terms of this sequence are called *numerals*—the 0th, 1st, $\ldots$, n^{th}, $\ldots$ *numeral*. The nth numeral is denoted by Δ_n. We can define Δ_n recursively by putting

$$\Delta_0 = 0, \text{ and } \Delta_{n+1} = S\Delta_n \text{ for every } n \in N.$$

In this way we have specified the meaning of the symbol Δ_n introduced in II.2.

We shall write

$$\alpha \leq \beta$$

instead of

$$Vw(w + \alpha = \beta)$$

where α and β are arbitrary terms and w is a variable which occurs in neither α nor β. (Regarding the way in which w may be fixed, compare a remark at the end of II.1.)

The axiom system of R contains all sentences which are particular

---

[9] Cf. [9], vol. 1, pp. 286 and 371.

instances of the following five axiom schemes where n and p are arbitrary natural numbers:

$\Omega_1.$ $\qquad\qquad\qquad \varDelta_n + \varDelta_p = \varDelta_{n+p}.$

$\Omega_2.$ $\qquad\qquad\qquad \varDelta_n \cdot \varDelta_p = \varDelta_{n \cdot p}.$

$\Omega_3.$ $\qquad\qquad\qquad \varDelta_n \neq \varDelta_p$ for $n \neq p.$

$\Omega_4.$ $\qquad x \leq \varDelta_n \to x = \varDelta_0 \lor x = \varDelta_1 \lor \ldots \lor x = \varDelta_n.$

$\Omega_5.$ $\qquad\qquad\qquad x \leq \varDelta_n \lor \varDelta_n \leq x.$

THEOREM 5. *In the sequence* N, P, Q, R *each of the theories* (*except* N) *is a subtheory of the preceding one.*

PROOF: P is obviously a subtheory of N. To show that Q is a subtheory of P it suffices to notice that Θ_3 is logically equivalent with the particular instance of the induction scheme $\varPi_3$ obtained by taking for $\varPhi$ the formula

$$u \neq 0 \to \lor y(u = \mathbf{S}y).$$

We shall now derive all the sentences Ω_1—Ω_5 from the axioms of Q.

The sentences Ω_1, i.e.,

(1) $\qquad\qquad \varDelta_n + \varDelta_p = \varDelta_{n+p}$ for all $n, p \in N$

are shown to be valid in Q by induction on p. (All the inductive arguments in this proof are of course metamathematical inductions, and not inductions within Theory Q.) In fact, for $p = 0$, (1) is a particular instance of Θ_4; when passing from p to $p + 1$, we apply Θ_5.

Similarly, by an induction on p based upon Θ_6, Θ_7, and (1), we derive the sentences Ω_2:

(2) $\qquad\qquad \varDelta_n \cdot \varDelta_p = \varDelta_{n \cdot p}$ for all $n, p \in N.$

Consider now the sentences Ω_3:

(3) $\qquad\qquad \varDelta_n \neq \varDelta_p$ for all $n, p \in N$ with $n \neq p.$

We can clearly assume that $n < p$. For $n = 0$ (and every $p > n$)

(3) is a particular case of Θ_2; when passing from n to $n + 1$ we apply Θ_1.

Θ_3 and Θ_5 imply

$$z + x = 0 \wedge x \neq 0 \to \mathsf{V}y[x = \mathsf{S}y \wedge \mathsf{S}(z + y) = 0];$$

hence, by Θ_2,

$$(4) \qquad\qquad\qquad x \leq \varDelta_0 \to x = \varDelta_0.$$

Similarly, for every natural number n we have by Θ_3 and Θ_5:

$$z + x = \varDelta_{n+1} \wedge x \neq 0 \to \mathsf{V}y[x = \mathsf{S}y \wedge \mathsf{S}(z + y) = \mathsf{S}\varDelta_n];$$

hence, by Θ_1,

$$(5) \qquad\qquad x \leq \varDelta_{n+1} \to x = 0 \vee \mathsf{V}y(x = \mathsf{S}y \wedge y \leq \varDelta_n).$$

From (4) and (5), by an induction with respect to n, we derive the sentences Ω_4:

$$(6) \quad x \leq \varDelta_n \to x = \varDelta_0 \vee x = \varDelta_1 \vee \ldots \vee x = \varDelta_n \text{ for every } n \in N.$$

(1) and (6) imply

$$(7) \qquad \varDelta_n \leq \varDelta_{n+1} \wedge (x \leq \varDelta_n \to x \leq \varDelta_{n+1}) \text{ for every } n \in N.$$

By an induction on n based upon Θ_4 and Θ_5 we obtain for any given natural number n

$$\mathsf{S}x + \varDelta_n = x + \varDelta_{n+1}.$$

Therefore, by Θ_3,

$$z + \varDelta_n = x \wedge z \neq 0 \to \mathsf{V}y(y + \varDelta_{n+1} = x);$$

and hence, with the help of (1) with $n = 0$,

$$(8) \qquad \varDelta_n \leq x \to x = \varDelta_n \vee \varDelta_{n+1} \leq x \text{ for every } n \in N.$$

We now can derive the sentences Ω_5:

$$(9) \qquad\qquad x \leq \varDelta_n \vee \varDelta_n \leq x \text{ for every } n \in N.$$

In fact, (9) for $n = 0$ follows from Θ_4; the passage from n to $n + 1$ is based upon (7) and (8).

In view of (1), (2), (3), (6), and (9), all the axioms of R turn out to be derivable from the axioms of Q, and the proof is complete.

In connection with this proof it may be interesting to notice that various simple sentences which are valid in N and in P are not derivable from the axioms of Q; e.g., the sentences

$$x \neq Sx, \quad 0 + x = x, \quad x \leq x, \quad 0 \cdot x = 0.$$

To show this, consider the system $\mathfrak{M} = \langle U, 0, S, +, \cdot \rangle$ constructed in the following way. U is the set consisting of all natural numbers and two additional elements, ∞_0 and ∞_1. 0 has its usual meaning; similarly, S, $+$, and $\cdot$ when applied to natural numbers. We put:

$S\infty_i = \infty_i$ (here and below $i = 0, 1$);

$\infty_i + n = \infty_i$ for every $n \in N$, $x + \infty_i = \infty_{1-i}$ for every $x \in U$;

$n \cdot \infty_i = \infty_i$ for every $n \in N$, $\infty_i \cdot 0 = 0$, $\infty_i \cdot x = \infty_{1-i}$ for every $x \in U$, $x \neq 0$.

It can easily be checked that all the axioms of Q are satisfied in $\mathfrak{M}$ so that $\mathfrak{M}$ is a model of Q; on the other hand, none of the four arithmetical sentences listed above is satisfied in $\mathfrak{M}$.

While Theory N is known not to be axiomatizable (see the remark following Theorem 9 in II.5), Theories P, Q, and R are obviously axiomatizable. Moreover, Q is finitely axiomatizable. On the other hand, it can easily be shown that R is not finitely axiomatizable; this reduces to showing that there is no finite set of axioms of R from which all the remaining axioms of R could be derived. The analogous problem for P is much more difficult and has been solved only recently; it has turned out that P is not finitely axiomatizable. [10]

II.4. Recursiveness and definability in subtheories of arithmetic.

When discussing in this section definable functions and sets (in the sense of II.2), we shall assume that the notion of definability has been relativized to the sequence of numerals $\Delta_0, \Delta_1, \ldots, \Delta_n, \ldots$ introduced in II.3.

It is well known that a function is recursive if and only if it is definable in Peano's arithmetic P; the same applies to a recursive set of natural numbers. [11] *A fortiori*, all recursive functions and sets

---

[10] See [25].

[11] This is actually one of the equivalent definitions of recursiveness which can be found in the literature; cf. [6].

are definable in every extension of P. In the next theorem we shall see that this result can be extended to many other subtheories of N—in fact, to Theory R and all its extensions.

It should be recalled that a function F is said to be definable in a theory T if there is a formula Φ of T which defines F; i.e., satisfies conditions (i)—(iii) stated explicitly in II.2. We have also mentioned a weaker notion of definable functions obtained by omitting condition (iii). It turns out that, with regard to any theory T which is an extension of R, the two notions of definable functions coincide. In fact, if a formula Φ defines F in the weaker sense, i.e., satisfies conditions (i) and (ii), then the formula

$$\Phi(u,v) \wedge \wedge y[\Phi(u,y) \to v \leq y]$$

(where y is any variable different from u and v and not occurring in Φ) proves to define F in the stronger sense, i.e., to satisfy conditions (i)—(iii). It should also be recalled that conditions (i) and (iii) can be replaced by one condition: (i') for every $n \in N$, the sentence

$$\wedge v[\Phi(\varDelta_n, v) \leftrightarrow v = \varDelta_{Fn}]$$

is valid in T. Condition (i') will actually be used in the following discussion. As regards condition (ii), it was stated in II.2 that this condition follows from (i) and (iii), and hence can be omitted, under the assumption that all the sentences $\varDelta_n \neq \varDelta_p$ with $n,p \in N$ and $n \neq p$ are valid in T; this assumption is clearly satisfied in case T is an extension of R.

THEOREM 6. *Every recursive function is definable in Theory R and hence also in every extension of R. The same applies to every recursive set of natural numbers.*

PROOF: We need a few auxiliary notions. Let E be a special function defined for every natural number n by the condition: En is the excess of n over a square, i.e., the unique $p \in N$ such that, for some $m \in N$,

$$m^2 + p = n < (m + 1)^2.$$

Given any two functions G and H, let $G + H$ and GH be respectively the functions K and L defined for every $n \in N$ by the formulas:

$$Kn = Gn + Hn, \quad Ln = G(Hn);$$

let G^{-1} be the function determined by the conditions:

$G^{-1}n = p$ if p is the smallest natural number m such that $Gm = n$;

$G^{-1}n = 0$ in case no such number m exists.

The following characterization of recursive functions due to Julia Robinson [21] proves to be convenient for our purposes:

(1) A function is recursive if and only if it belongs to every set $\mathfrak{F}$ of functions satisfying the following conditions:

(i) $S, E \in \mathfrak{F}$;

(ii) if $G, H \in \mathfrak{F}$, then $G + H$, $GH \in \mathfrak{F}$;

(iii) if $G \in \mathfrak{F}$ and G assumes all natural numbers as values, then $G^{-1} \in \mathfrak{F}$.

We want to apply (1) by taking for $\mathfrak{F}$ the set of all functions definable in R. With this in view, we shall first show that this set actually satisfies conditions (i)–(iii) of (1).

(2) The function S is definable in R.

In fact, $\mathbf{S}u = v$ is clearly a defining formula for S.

(3) The function E is definable in R.

To prove (3) we shall show that the formula

(4) $\Phi = \bigvee x[x \leq u \wedge v \leq x + x \wedge u = (x \cdot x) + v]$

(where x is a variable different from u and v) defines E. This amounts to showing that, for any given $n \in N$, the sentences

(5) $$v = \Delta_{En} \rightarrow \Phi(\Delta_n, v),$$

(6) $$\Phi(\Delta_n, v) \rightarrow v = \Delta_{En}$$

are valid in R. By the definition of E, there is a natural m such that

$$m^2 + En = n < (m + 1)^2$$

and hence

$$m \leqslant n \text{ and } En \leqslant 2 \cdot m.$$

By Ω_1 and Ω_2, we conclude that the sentence

$$\Delta_m \leq \Delta_n \wedge \Delta_{En} \leq \Delta_m + \Delta_m \wedge \Delta_n = (\Delta_m \cdot \Delta_m) + \Delta_{En}$$

is valid in R; in view of (4), this clearly implies the validity of (5). On the other hand, from (4), Ω_1, Ω_2, and Ω_4 we obtain (as a valid formula of R)

$$\Phi(\Delta_n, v) \to (v \leq \Delta_{2\cdot 0} \wedge \Delta_n = \Delta_{0'} + v) \vee \ldots \vee (v \leq \Delta_{2\cdot n} \wedge \Delta_n = \Delta_{n'} + v).$$

Hence, by applying Ω_4 again, we derive a sentence of the form

(7) $\Phi(\Delta_n, v) \to \Psi$

where Ψ is a disjunction of formulas

$$v = \Delta_p \wedge \Delta_n = \Delta_{m'+p}$$

with $m \leq n$ and $p \leq 2 \cdot m$. En is uniquely determined as the natural number p such that $n = m^2 + p$ and $p \leq 2 \cdot m$ for some $m \in N$. Thus, if $p \neq En$, we have $n \neq m^2 + p$ whenever $p \leq 2 \cdot m$; therefore, by Ω_3, all the formulas

$$\Delta_n \neq \Delta_{m'+p}$$

with $p \neq En$ and $p \leq 2 \cdot m$ are valid in R. Hence, using (7), we finally conclude that (6) is valid. Consequently, Φ defines E.

Next we prove:

(8) If the functions G and H are definable in R, then so are $G + H$ and GH.

In fact, Φ and Ψ being two formulas respectively defining G and H, it is easily seen that the formula

$$\vee x \vee y [v = x + y \wedge \Phi(u,x) \wedge \Psi(u,y)]$$

defines $G + H$, while the formula

$$\vee z [\Psi(u,z) \wedge \Phi(z,v)]$$

defines GH.

(9) If the function G is definable in R and assumes all natural numbers as values, then G^{-1} is definable in R.

To prove this, we consider a formula Φ which defines G and we put

(10) $\Psi = \Phi(v,u) \wedge \wedge y [\Phi(y,u) \to v \leq y]$.

We shall show that Ψ defines G^{-1}, i.e., that the sentences

(11) $\Psi(\Delta_n, v) \to v = \Delta_{G^{-1}n}$

and

(12) $$v = \Delta_{G^{-1}n} \to \Psi(\Delta_n, v)$$

are valid in R for any given $n \in N$.

By the definition of G^{-1} we have $Gm \neq n$ for every $m < G^{-1}n$; hence, by Ω_3, all the sentences

$$\Delta_{Gm} \neq \Delta_n$$

with $m < G^{-1}n$ are valid. Since the sentences

(13) $$\Phi(\Delta_m, v) \leftrightarrow v = \Delta_{Gm}$$

are also valid for every m, we derive by Ω_4 the sentence

(14) $$\Phi(v, \Delta_n) \wedge v \leq \Delta_{G^{-1}n} \to v = \Delta_{G^{-1}n}.$$

Moreover, $G(G^{-1}n) = n$ (since G assumes every natural number as a value), and hence we obtain from (13)

(15) $$\Phi(\Delta_{G^{-1}n}, \Delta_n).$$

From (10) we see that the sentence

$$\Psi(\Delta_n, v) \to \Phi(v, \Delta_n) \wedge [\Phi(\Delta_{G^{-1}n}, \Delta_n) \to v \leq \Delta_{G^{-1}n}]$$

is valid; combined with (14) and (15), this sentence implies (11). On the other hand, from Ω_5 and (14) we obtain:

$$y \leq \Delta_{G^{-1}n} \vee \Delta_{G^{-1}n} \leq y,$$
$$\Phi(y, \Delta_n) \wedge y \leq \Delta_{G^{-1}n} \to y = \Delta_{G^{-1}n}.$$

Together with (15) and Ω_1, these sentences give

(16) $$\Phi(\Delta_{G^{-1}n}, \Delta_n) \wedge \wedge y[\Phi(y, \Delta_n) \to \Delta_{G^{-1}n} \leq y].$$

From (10) and (16) we derive

$$\Psi(\Delta_n, \Delta_{G^{-1}n}),$$

and this sentence obviously implies (12).

We have thus shown that Ψ defines G^{-1} and that consequently (9) holds. By (1), (2), (3), (8), and (9), every recursive function is definable in R and *a fortiori* in every extension T of R. If P is a recursive set of natural numbers, then its characteristic function C_P is also recursive and therefore definable in R (and T); as we know, this implies that the set P itself is definable. The proof is thus complete.

COROLLARY 7. T *being any consistent and axiomatizable extension of* R, *a function is definable in* T *if and only if it is recursive; the same applies to sets of natural numbers.*

PROOF: The corollary follows directly from Theorems 3 and 6.

COROLLARY 8. *If* T *is a consistent extension of* R, *then the set* V *is not definable in* T. [12]

PROOF: In II.2 we have defined the function D and the set V, relative to a fixed one-to-one correspondence between natural numbers and expressions of T. (We recall that V is the set of all natural n such that the correlated expression E_n is a sentence valid in T.) This correspondence is subjected to usual conditions of recursiveness, and consequently the function D proves to be recursive. Therefore, by Theorem 6, D is definable in T, and hence, by Theorem 1, V is not definable.

II.5. **Undecidability of subtheories of arithmetic.** By combining the results obtained in II.2 and II.4 we can now establish the fundamental results of this paper: the essential undecidability of those subtheories of N which were described in II.3 and the undecidability of all subtheories of N.

THEOREM 9. *Theory* R *is essentially undecidable. The same applies to every consistent extension of* R *and, in particular, to* Q, P, *and* N. [13]

---

[12] In its application to a narrower class of theories, in fact to those theories which are consistent extensions of P, this result coincides with Theorem 1 in [15], p. 88; in its application to Theory N it coincides with Theorem I in [31], p. 370.

[13] This result seems to be essentially new as far as Theories R and Q are concerned. On the other hand, the essential undecidability of P was

PROOF: Since all the theories mentioned in this theorem are consistent, the theorem is an immediate consequence of Corollary 2 and Theorem 6.

As was pointed out in I.3, every theory which is not axiomatizable is automatically undecidable. The converse, of course, does not hold; by Theorem 9, P, Q, and R are instances of theories which are both axiomatizable and essentially undecidable. On the other hand, it is known that no complete and undecidable theory is axiomatizable; see, e.g., Theorem 1 of I.3. Theory N is complete (since the set of all valid sentences of N coincides with the set of sentences which are satisfied in a single model). Hence, again by Theorem 9, N is not axiomatizable. We can reach this conclusion also in a different way. There are sets which are definable in N, but are not recursive. This follows, e.g., from the fact that all recursively enumerable sets are known to be definable in N. More specifically, the set V corresponding to Theory P is known to be definable in N [14]; this set, however, is not recursive since P is undecidable. Hence N is not axiomatizable by Theorem 3.

The importance and strength of Theorem 9 results from the fact that it implies the essential undecidability of a subtheory of N which is finitely axiomatizable, in fact, of Theory Q; cf. I.6. Keeping this in mind, we obtain at once

---

established by Rosser in [24]; in fact, P was the first axiomatic theory which was shown to be essentially undecidable. It may be interesting to notice that, in establishing his result, Rosser used a refined and more complicated variant of an argument whose idea goes back to Gödel [7] and which was applied in Church [3] to obtain the simple undecidability of P; on the other hand, the argument used in this paper seems to be even simpler than the original arguments of Gödel and Church. As regards N, since this theory is complete, its essential undecidability reduces to simple undecidability (cf., e.g., Theorem 1 of I.3); and if one identifies recursive sets and functions with those which are definable in P (see footnote 11), then the undecidability of N becomes a direct consequence of the fact that the set of all valid sentences of N is not definable in N —a fact established by Tarski in [31] (pp. 370 ff. and in particular footnotes 88 and 95).

[14] See, e.g., [15], pp. 88 and 91.

CONCLUSION CRITICAL... COROLLARY 10. *Every subtheory of* N *having the same constants as* N *is undecidable.* [15] *The same applies to every theory which is compatible with* Q *and contains all constants occurring in* Q.

PROOF: By Theorem 9 of this paper and Theorem 6 of I.3.

We could try to improve that part of Theorem 9 which refers to Theory Q, by constructing some finitely axiomatizable subtheories of Q which would be still essentially undecidable. The simplest way of constructing subtheories of Q consists of course in omitting some of the axioms. In the next theorem we shall show, however, that none of the theories thus obtained is essentially undecidable (though, by Corollary 10, all of them are undecidable).

THEOREM 11. *No axiomatic subtheory of* Q *obtained by removing any one of the seven axioms from the axiom system is essentially undecidable.*

PROOF: Let Q_n be the axiomatic subtheory of Q obtained by omitting Axiom Θ_n, $n = 1, 2, \ldots, 7$. We shall prove that none of these theories is essentially undecidable by constructing, for each of them, a consistent and decidable extension T_n. With the exception of T_6, each of the theories T_n will have the same constants as Q. The consistency of the theories T_n will obviously follow from their descriptions; also it will be very easy to check that T_n is an extension of Q_n, i.e., that all the axioms of Q_n are valid in T_n. Thus our task reduces to showing that each of the theories T_n is actually decidable.

T_1 is the theory determined by the following stipulation: a sentence is valid in T_1 if it is satisfied in the system $\mathfrak{M} = \langle U, 0, S, +, \cdot \rangle$ where

(i) U is the set consisting of two integers, 0 and 1;
(ii) 0 and $\cdot$ have their ordinary meaning;
(iii) $S0 = S1 = 1$;

---

[15] Only special cases of this general result are known from the literature. Thus, as regards P and N itself, cf. footnote 13 above; furthermore, the results of Church in [2] imply that the subtheory of N which has the same constants as N but has no non-logical axioms (i.e., the first-order predicate logic with constants restricted to those of N) is undecidable.

(iv) $x + y = 0$ if $x = y = 0$, $x + y = 1$ if $x,y \in U$ and $x = 1$ or $y = 1$.

The decidability of T_1 is simply a consequence of the fact that the universe U of the model $\mathfrak{M}$ is finite. A decision procedure for T_1 can be roughly described as follows. [16] Φ being any formula (in which the variable u does not occur bound), the sentences

$$\wedge u \Phi(u) \leftrightarrow \Phi(0) \wedge \Phi(\mathsf{S}0),$$

$$\vee u \Phi(u) \leftrightarrow \Phi(0) \vee \Phi(\mathsf{S}0)$$

are easily seen to be valid in T_1. Hence, with every sentence Φ we can correlate a sentence Φ^* containing no variables and no quantifiers such that Φ is valid if and only if Φ^* is valid; and this correlation between Φ and Φ^* is recursive. Consequently, it suffices to indicate a decision procedure for sentences containing no variables. For atomic sentences of this type, i.e., for equations between terms, the procedure is based directly upon the definitions of operations which occur in the model $\mathfrak{M}$; for compound sentences the procedure reduces to the familiar decision procedure for the ordinary sentential calculus.

A sentence is valid in T_2 if it is satisfied in the system $\langle U,\ 0,\ S,\ +,\ \cdot \rangle$ where U is the set consisting of the single element 0 and where consequently $S0 = 0 + 0 = 0 \cdot 0 = 0$. Again, T_2 is decidable because the set U is finite; and in establishing a decision procedure for T_2 we make use of the fact that, for every formula Φ, the sentences

$$\wedge u \Phi(u) \leftrightarrow \Phi(0) \text{ and } \vee u \Phi(u) \leftrightarrow \Phi(0)$$

are valid in T_2.

A sentence is valid in T_3 if it is satisfied in the system $\langle P,\ 0,\ S,\ +,\ \cdot \rangle$ where P is the set of all non-negative real numbers, $Sx = x + 1$ for every $x \in P$, and 0, $+$, $\cdot$ have their ordinary meaning. Let $\bar{\mathsf{T}}$ be a theory with five constants, $\mathbf{P}$, $\mathbf{0}$, $\mathbf{S}$, $\boldsymbol{+}$, $\cdot$, where $\mathbf{P}$ is a unary predicate; by definition, a sentence is valid in

---

[16] The method applied here is a simple particular case of the general method of eliminating quantifiers; compare (also for historical references) [30], p. 15 and note 11.

$\overline{T}$ if it is satisfied in the system $\langle R, P, 0, S, +, \cdot \rangle$ where R is the set of all real numbers, and the remaining symbols have the same meanings as in T_3. $\overline{T}$ is decidable; this we easily derive from the results in [30] if we remember that non-negative real numbers coincide with squares of arbitrary real numbers. With every sentence Φ of T_3 we correlate the relativized sentence $\Phi^{(P)}$; cf. I.5. As is easily seen, Φ is valid in T_3 if and only if $\Phi^{(P)}$ is valid in $\overline{T}$. Hence a decision procedure for $\overline{T}$ automatically yields a decision procedure for T_3, and T_3 proves to be decidable.

A sentence is valid in T_4 if it is satisfied in the system $\mathfrak{M} = \langle N, 0, S, +, \cdot \rangle$ where $N, 0, S$ have the usual meaning while the operations $+$ and $\cdot$ on $N \times N$ to N are defined by the formulas:

$$n + p = p, \ n \cdot 0 = 0, \ n \cdot p = n \text{ for } p \neq 0.$$

Let T^* be a theory with only two constants $\mathbf{0}$ and $\mathbf{S}$; a sentence of T^* is valid in T^* if and only if it is valid in T_4, i.e., is satisfied in the system $\langle N, 0, S \rangle$ where $N, 0, S$ have their usual meaning. The sentences

$$y = x + y,$$
$$x = y \cdot z \leftrightarrow (x = \mathbf{0} \wedge z = \mathbf{0}) \vee (x = y \wedge z \neq \mathbf{0})$$

are clearly valid in T_4. Using these sentences, with every sentence Φ of T_4 we can recursively correlate a sentence Φ^* of T^* such that $\Phi \leftrightarrow \Phi^*$ is valid in T_4, and consequently Φ is valid in T_4 if and only if Φ^* is valid in T^*. Hence the decision problem for T_4 reduces to that for T^*. T^* is known to be decidable (this follows, e.g., from the results in [17]), and therefore T_4 is also decidable.

T_5 differs from T_4 only in the definitions of the operations $+$ and $\cdot$ which occur in the model $\mathfrak{M}$; we put

$$n + p = n \text{ and } n \cdot p = 0 \text{ for all } n,p \in N.$$

The proof of decidability remains unchanged.

T_6 is a theory with seven constants: $\mathbf{0}, \mathbf{S}, \boldsymbol{+}, \cdot, \dot{\mathbf{0}}, \dot{\mathbf{S}}, \dot{\boldsymbol{+}}$. A sentence is valid in T_6 if it is satisfied in the system $\langle N, 1, S', \oplus, \cdot, 0, S, + \rangle$ where $N, 1, 0, S, +$ have their usual meaning while $S', \oplus$, and $\cdot$ are defined by the conditions

$$S'0 = 0, \; S'n = Sn \text{ if } n \neq 0;$$
$$n \oplus p = 0 \text{ if } n = 0 \text{ or } p = 0; \; n \oplus p = n + p - 1 \text{ if } n \neq 0 \text{ and } p \neq 0;$$
$$n \cdot p = 0 \text{ for all } n, p \in N.$$

Let T^+ be a theory with three constants: $\dot{0}$, $\dot{S}$, $\dot{+}$; a sentence of T^+ is valid in T^+ if it is valid in T_6, i.e., is satisfied in the system $\langle N, 0, S, + \rangle$ where all the symbols have their usual meaning. The following sentences are clearly valid in T_6:

$$0 = \dot{S}\dot{0},$$
$$x = Sy \leftrightarrow (x = \dot{0} \wedge y = \dot{0}) \vee (x = \dot{S}y \wedge y \neq \dot{0}),$$
$$x = y \dotplus z \leftrightarrow (x = \dot{0} \wedge y = \dot{0}) \vee (x = \dot{0} \wedge z = \dot{0}) \vee$$
$$(\dot{S}x = y \dotplus z \wedge y \neq \dot{0} \wedge z \neq \dot{0}),$$
$$y \cdot z = \dot{0}.$$

With the help of these sentences, arguing as in the case of T_4, we reduce the decision problem for T_6 to that for T^+. The latter theory is known to be decidable (see [17]); hence T_6 is also decidable.

(Another, perhaps more natural, example of a decidable extension of Q_6 is provided by a theory T_6' which has the same constants as Q. A sentence is valid in T_6' if it is satisfied in the system $\langle U, 0, S, +, \cdot \rangle$ where (i) U is the set consisting of the natural numbers and an additional element ∞; (ii) $0, S, +$ have their usual meaning when applied to natural numbers, and in addition

$$S\infty = x + \infty = \infty + x = \infty$$

for every $x \in U$; (iii) $x \cdot y = \infty$ for all $x, y \in U$. The decision problem for T_6' can again be reduced to that for T^+; however, the reduction is somewhat more involved than in the case of T_6.)

A sentence is valid in T_7 if it is satisfied in the system $\langle N, 0, S, +, \cdot \rangle$ where $n \cdot p = 0$ for all $n, p \in N$. We see at once that the decision problem for the theory T_7 again reduces to that for the theory T^+ described above in connection with T_6. (The symbols $\dot{0}$, $\dot{S}$, $\dot{+}$ are now replaced in T^+ by 0, S, $+$.) Hence T_7 is decidable, and the proof is complete.

II.6. Extension of the results to other arithmetical theories and to various theories of rings. The results obtained in the preceding

sections and in particular in II.5 can be carried over to formalized systems of the arithmetic of natural numbers with different sets of constants. Consider, e.g., the theory N* which differs from N only in that the unary operation symbol S has been replaced by the individual constant 1 denoting the integer one; the definition of validity in N* is entirely analogous to that in N. We can construct axiomatic subtheories P*, Q*, R* of N* closely related to the subtheories of N described in II.3—simply by replacing in the axioms of P, Q, R all the terms of the form $S\alpha$ by $\alpha + 1$. It is easily seen that Theorem 9 and Corollary 10 remain valid if we replace in them N, P, Q, R by N*, P*, Q*, R*, respectively. To prove this, it suffices to show that each of the theories N, P, ... is interpretable in the corresponding theory N*, P*, ... in the sense of I.4, and then to apply Theorems 7 and 8 of I.4. For instance, if we add the symbol S to the constants of Theory Q* and the sentence

$$x = Sy \leftrightarrow x = y + 1$$

to its axioms, we clearly obtain an extension of Q; and since the new axiom is a possible definition of S in Q*, Q proves to be indeed interpretable in Q*. Incidentally, it is easily seen that, conversely, Q* is interpretable in Q. On the other hand, it may be interesting to notice that Theorem 11 partly fails when applied to Q*. In fact, by removing Θ_4^* from the axiom system of Q* we obtain a subtheory of Q* in which Q* proves to be interpretable and which therefore is essentially undecidable. If, however, we omit any of the remaining six axioms of Q*, then the resulting subtheory will no longer be essentially undecidable.

Instead of replacing S by 1, we can eliminate S as well as 0 from the list of constants of N altogether. The system N+ of the arithmetic of natural numbers thus obtained contains $+$ and $\cdot$ as the only constants. N is interpretable in N+ since every valid sentence of N is derivable from the set of valid sentences of N+ supplemented, e.g., by the following two possible definitions of 0 and S:

$$x = 0 \leftrightarrow x = x + x,$$
$$x = Sy \leftrightarrow \forall z(x = y + z \wedge z \neq z + z \wedge z = z \cdot z).$$

Therefore Q is also interpetable in N^+, by Theorems 8(i) and 7(ii) of I.4, N^+ and all its subtheories are undecidable, and there is a finitely axiomatizable subtheory Q^+ of N^+ which is essentially undecidable. It is easy actually to construct such a subtheory Q^+ by appropriately transforming the axioms of Q.

When we analyze the arguments in the preceding sections and specifically the proof of Theorem 6, we notice that all our results remain valid if, instead of Theory R, we use a weaker theory R' in which the axiom scheme Ω_2 is replaced by a more special scheme:

Ω_2'. $\Delta_n \cdot \Delta_n = \Delta_{n^2}$ for every natural n.

Making use of this observation, we arrive at the conclusion that in all the theories discussed above the symbol $\cdot$ can be replaced by the unary operation symbol K denoting the operation of squaring a number. In particular we obtain a finitely axiomatizable and essentially undecidable theory Q' by replacing Axioms Θ_6 and Θ_7 by the following two axioms (which constitute together a recursive definition of K):

Θ_6'. $K0 = 0$.

Θ_7'. $K(Sx) = Kx + S(x + x)$. [17]

The fundamental results of II.5 can be extended to the *arithmetic of positive integers* and to that of arbitrary integers. As a simple example of a finitely axiomatizable and essentially undecidable subtheory of the arithmetic of positive integers we may mention a theory $Q^\times$ closely related to Q and Q*; it contains 1, $+$, and $\cdot$ as the only constants, and is based upon the following six axioms:

$\Theta_1^\times$. $x + 1 = y + 1 \rightarrow x = y$.

$\Theta_2^\times$. $1 \neq y + 1$.

$\Theta_3^\times$. $x \neq 1 \rightarrow \vee y(x = y + 1)$.

$\Theta_4^\times$. $x + (y + 1) = (x + y) + 1$.

$\Theta_5^\times$. $x \cdot 1 = x$.

$\Theta_6^\times$. $x \cdot (y + 1) = (x \cdot y) + x$.

---

[17] For the extension of these results to formalized theories of natural numbers with still other sets of constants see [20], pp. 112 f.

As regards the *arithmetic of arbitrary integers*, we shall consider two formalized systems of this arithmetic, J and $J^<$. The only constants occurring in J are $+$ and $\cdot$; $J^<$ contains in addition the binary predicate $<$ (to denote the less-than relation). A sentence is valid in J if it is satisfied in the system $\langle I, +, \cdot \rangle$ where all the symbols have their usual meaning; similarly for $J^<$. We have the following

THEOREM 12. *Theory J and all its subtheories (having the same constants as J) are undecidable; there are finitely axiomatizable subtheories of J which are essentially undecidable. The same applies to $J^<$.*

PROOF: Let **N** be any unary predicate, and let $Q^{(N)}$ be the theory obtained from Q by relativizing the quantifiers to **N**; see I.5. By Theorems 9 (ii) and 10 of I and by Theorem 9 of the present paper,

(1) $Q^{(N)}$ is finitely axiomatizable and essentially undecidable.

We want to show that

(2) $Q^{(N)}$ is interpretable in J.

To prove this, consider the formalized arithmetic $\overline{J}$ of integers with five constants **N, 0, S, $+$, $\cdot$**; a sentence is valid in $\overline{J}$ if and only if it holds in the system $\langle I, N, 0, S, +, \cdot \rangle$ where all symbols have their usual meaning. $\overline{J}$ is obviously an extension of J. Every sentence Φ which is valid in Q is also valid in the arithmetic N of natural numbers, and hence the relativized sentence $\Phi^{(N)}$ is valid in $\overline{J}$; thus $\overline{J}$ is an extension of $Q^{(N)}$ as well. The following three sentences are clearly valid in $\overline{J}$:

(3) $Nx \leftrightarrow \vee y \vee z \vee u \vee v [x = (y \cdot y) + (z \cdot z) + (u \cdot u) + (v \cdot v)]$,
(4) $x = 0 \leftrightarrow x = x + x$,
(5) $x = Sy \leftrightarrow \vee z (x = y + z \wedge z \neq z + z \wedge z = z \cdot z)$.

In particular, (3) is a formulation of the well-known theorem by which natural numbers coincide with those integers which are sums of four squares. Obviously, (3) is a possible definition of **N** in J; (4) is a possible definition of **0** since the sentence

$$\vee x [x = x + x \wedge \wedge y (y = y + y \to x = y)]$$

is known to be valid in J (cf. I.4); and for a similar reason (5) is a possible definition of **S**. With every valid sentence Ψ of $\overline{\mathsf{J}}$ we correlate a sentence Ψ^+ of J which is obtained from Ψ by eliminating the constants **N**, **0**, **S** with help of (3)–(5). It is easily seen that Ψ^+ is valid in J and that Ψ is derivable from the set consisting of the sentences Ψ^+, (3), (4), (5). Hence every valid sentence of $\overline{\mathsf{J}}$ is derivable from the set of all valid sentences of J supplemented by the sentences (3)–(5). Thus, according to the remarks in I.4, $Q^{(N)}$ is indeed interpretable in J.

From (1) and (2), by applying Theorems 8(i) and 7(ii) of I.4, we obtain immediately that part of our theorem which concerns J. Thus, in particular, it turns out that there are finitely axiomatizable subtheories of J which are essentially undecidable. If we wish actually to construct such a subtheory of J, we recall that a finite axiom system for Q is available; hence, by analyzing the proof of Theorem 9 in I.5, we obtain a finite axiom system for the relativized theory $Q^{(N)}$. Using sentences (3), (4), (5) stated above, we eliminate the constants **N**, **0**, **S** from the axioms of $Q^{(N)}$, and we arrive at a set A of sentences which are valid in J. The subtheory S of J based upon the set A as an axiom system is both finitely axiomatizable and essentially undecidable (since A is finite and $Q^{(N)}$ is clearly interpretable in S).

The proof for $\mathsf{J}^<$ is entirely analogous; we can choose in this case a simpler sentence as a possible definition of **N**, e.g.,

$$\mathbf{N}x \leftrightarrow x = x + x \lor x < x + x.$$

Among axiomatic subtheories of J and $\mathsf{J}^<$ we find several which are of a special interest from the point of view of modern algebra. All these subtheories are undecidable by Theorem 12; some of them present rather simple and interesting examples of finitely axiomatizable theories which are essentially undecidable.

To describe these theories consider the following sentences:

$$\Sigma_1. \qquad\qquad x + (y + z) = (x + y) + z.$$
$$\Sigma_2. \qquad\qquad x + y = y + x.$$

$\Sigma_3.$ $\lor z(x = y + z).$

$\Sigma_4.$ $x \cdot (y \cdot z) = (x \cdot y) \cdot z.$

$\Sigma_5.$ $x \cdot (y + z) = (x \cdot y) + (x \cdot z).$

$\Sigma_6.$ $(y + z) \cdot x = (y \cdot x) + (z \cdot x).$

$\Sigma_7.$ $x \cdot y = y \cdot x.$

$\Sigma_8.$ $z = z + z \land z = x \cdot y \to z = x \lor z = y.$

$\Sigma_9.$ $\lor u[u \neq u + u \land \land x(x \cdot u = x \land u \cdot x = x)].$

$\Sigma_1^<.$ $\sim (x < x).$

$\Sigma_2^<.$ $x \neq y \to x < y \lor y < x.$

$\Sigma_3^<.$ $x < y \land y < z \to x < z.$

$\Sigma_4^<.$ $y < z \to x + y < x + z.$

$\Sigma_5^<.$ $z = z + z \land z < x \land z < y \to z < x \cdot y.$

$\Sigma_6^<.$ $\lor x \lor y[x < y \land \sim \lor z(x < z \land z < y)].$

The axiomatic theory, with $+$ and $\cdot$ as the only constants, whose axiom system consists of $\Sigma_1 - \Sigma_6$ is called the *elementary theory of rings*; clearly, a sentence is valid in this theory if and only if it is satisfied in every system $\Re = \langle R, +, \cdot \rangle$ which is a ring in the sense of modern algebra. If we add Σ_7, or Σ_7 and Σ_8, to the axioms of the theory of rings, we obtain the *elementary theory of commutative rings*, or *of integral domains*, respectively. By including $<$ in the constants and $\Sigma_1^< - \Sigma_5^<$ in the axiom systems of these theories, we arrive at the *elementary theories of ordered rings, ordered commutative rings*, and *ordered integral domains*; actually, the latter two theories prove to coincide. If, in turn, we supplement the axiom systems with $\Sigma_6^<$, the resulting theories are referred to as the *elementary theories of non-densely ordered rings* and *non-densely ordered commutative rings*. Finally, we can include Σ_9 in any of the axiom systems mentioned above, and we obtain various theories *with unit*, e.g., the *elementary theory of rings with unit*.

Sentences $\Sigma_1 - \Sigma_9$ are obviously valid both in J and $J^<$; sentences $\Sigma_1^< - \Sigma_6^<$ are valid in $J^<$. Hence, as a direct consequence of Theorem 12, we obtain

COROLLARY 13. *The elementary theories of rings, commutative rings, integral domains, ordered rings, and ordered commutative rings, without or with unit, are undecidable.* [18]

The theories mentioned in Corollary 13 are not essentially undecidable; for a consistent extension of these theories, in fact, the elementary theory of the ordered field of real numbers, has been shown to be decidable (see [30]). On the other hand, we have

THEOREM 14. *The elementary theories of non-densely ordered rings and non-densely ordered commutative rings, without or with unit, are essentially undecidable.*

PROOF: It suffices to show that the elementary theory T of non-densely ordered rings without unit, i.e., the theory based upon Axioms Σ_1–Σ_6 and $\Sigma_1^<$–$\Sigma_6^<$, is essentially undecidable; this automatically implies the essential undecidability of the remaining theories mentioned in the theorem since each of them is an extension of T.

We consider the relativized theory $\mathsf{Q}^{(N)}$ involved in the proof of Theorem 12; however, we replace in it the symbol $\cdot$ by $\times$. As we know, $\mathsf{Q}^{(N)}$ is finitely axiomatizable and essentially undecidable. The following sentences form an adequate axiom system for $\mathsf{Q}^{(N)}$ (cf. the proof of Theorem 9 in I.5):

(1) $\qquad\qquad Nx \wedge Ny \wedge Sx = Sy \to x = y.$

(2) $\qquad\qquad\qquad Ny \to 0 \neq Sy.$

(3) $\qquad\qquad Nx \wedge x \neq 0 \to \vee y(Ny \wedge x = Sy).$

(4) $\qquad\qquad\qquad Nx \to x + 0 = x.$

(5) $\qquad\qquad Nx \wedge Ny \to x + Sy = S(x + y).$

(6) $\qquad\qquad\qquad Nx \to x \times 0 = 0.$

(7) $\qquad\qquad Nx \wedge Ny \to x \times Sy = (x \times y) + x.$

(8) $\qquad\qquad\qquad\qquad N0.$

---

[18] For related results see [23] (the undecidability of the elementary theories of various special rings) and [20], p. 113 (the undecidability of the elementary theory of fields).

(9) $\mathbf{N}x \to \mathbf{N}(\mathbf{S}x)$.

(10) $\mathbf{N}x \wedge \mathbf{N}y \to \mathbf{N}(x + y)$.

(11) $\mathbf{N}x \wedge \mathbf{N}y \to \mathbf{N}(x \times y)$.

We want to show that $Q^{(\mathbf{N})}$ is interpretable in T. In order to prove this, following the remarks in I.4, we construct an extension $\overline{\mathsf{T}}$ of T by adding $\mathbf{N}, \mathbf{0}, \mathbf{S}, \times$ to the set of constants and by adjoining the following four sentences to the axiom system of T:

(12) $\mathbf{N}u \leftrightarrow (u = u + u \vee u < u + u) \wedge \forall t[t < t + t \wedge$
$\sim \forall x(x < t \wedge x < x + x) \wedge u \cdot t = t \cdot u \wedge \bigwedge y \forall z(y \cdot u = z \cdot t)]$.

(13) $u = \mathbf{0} \leftrightarrow u = u + u$.

(14) $u = \mathbf{S}v \leftrightarrow \forall t[t < t + t \wedge \sim \forall x(x < t \wedge x < x + x) \wedge$
$u = v + t]$.

(15) $u = v \times w \leftrightarrow \forall t\{t < t + t \wedge \sim \forall x(x < t \wedge x < x + x) \wedge$
$[u \cdot t = v \cdot w \vee [u = v \wedge \bigwedge y(y \cdot t \neq v \cdot w)]]\}$.

(12) is obviously a possible definition of $\mathbf{N}$ in T. We have to show that (13)–(15) are possible definitions of $\mathbf{0}, \mathbf{S}$, and $\times$; for instance, denoting by Φ the right side of the equivalence (14), we have to show that the following sentence is derivable from the axioms of T (see I.4):

$$\bigwedge v \forall u\{\Phi(u,v) \wedge \bigwedge w[\Phi(w,v) \to u = w]\}.$$

We have also to show that $\overline{\mathsf{T}}$ is an extension of $Q^{(\mathbf{N})}$, i.e., that all the sentences (1)–(11) are derivable from the axiom system of T supplemented by sentences (12)–(15). Instead of carrying through the derivations in a formal way, we shall outline the proof in ordinary mathematical language.

Let $\mathfrak{R} = \langle R, +, \cdot, < \rangle$ be any model of T, i.e., any non-densely ordered ring. As is well known the zero element z of this ring is uniquely characterized by the formula $z = z + z$ (hence (13) is a possible definition of $\mathbf{0}$ in T); we shall denote this element by 0.

Since $\mathfrak{R}$ is non-densely ordered, there are two elements $x, y \in R$ such that $x < y$ and that there is no element $z \in R$ with $x < z < y$. Let t be the element of R such that $x + t = y$. This element t has the following two properties which determine it uniquely: (i) t is a positive element, i.e., $0 < t$ (or, what amounts to the same, $t < t + t$); (ii) there is no positive element v with $v < t$, for otherwise we would have $x < x + v < y$. We put $Su = u + t$ for every $u \in R$. (From the above it follows that (14) is a possible definition of S in T.)

In every ordered ring the formula $x \cdot z = y \cdot z$ with $z \neq 0$ implies $x = y$. Hence we can correlate an element $v \times w$ with any couple of elements $v, w \in R$ in the following way: if $v \cdot w = u \cdot t$ for some $u \in R$ (t having the same meaning as before), we put $v \times w = u$; if $v \cdot w \neq y \cdot t$ for every $y \in R$, we put $v \times w = v$. Thus $\times$ is a uniquely determined binary operation on $R \times R$ to R (and (15) is a possible definition of $\times$ in. T).

Let N' be the set of all non-negative elements $u \in R$ ($u = u + u$ or $u < u + u$) such that $u \cdot t = t \cdot u$ and that, for every $y \in R$, there is a $z \in R$ with $y \cdot u = z \cdot t$. We shall show that $\langle N', 0, S, +, \times \rangle$ is a model of $Q^{(\mathbf{N})}$, i.e., that it satisfies sentences (1)–(11).

It obviously follows from general properties of rings and from the definition of S that sentences (1), (2), (4), (5) are satisfied. We have $0 \in N'$ since $0 \cdot t = t \cdot 0 = y \cdot 0$ for every $y \in R$. Obviously $t \in N'$. If $x, y \in N'$, then $x + y$ is clearly non-negative; $(x + y) \cdot t = (x \cdot t) + (y \cdot t) = (t \cdot x) + (t \cdot y) = t \cdot (x + y)$; for every $u \in R$ there are $z', z'' \in R$ such that $u \cdot (x + y) = (u \cdot x) + (u \cdot y) = (z' \cdot t) + (z'' \cdot t) = (z' + z'') \cdot t$; thus, finally, $x + y \in N'$. In particular, if $x \in N'$, then $Sx = x + t \in N'$. Hence, sentences (8), (9), and (10) are satisfied.

If $x \in N'$ and $x \neq 0$, then x is positive and hence, by the definition of t, $t = x$ or $t < x$. Consequently there is a non-negative element $y \in R$ such that $x = y + t = Sy$. We have $(y + t) \cdot t = x \cdot t = t \cdot x = t \cdot (y + t)$ and therefore $y \cdot t = t \cdot y$; similarly we show that for every $u \in R$ there is a $z \in R$ with $u \cdot y = z \cdot t$. Thus $y \in N'$, and sentence (3) holds.

For every x, $x \cdot 0 = 0 \cdot t$. Hence, by the definition of $\times$, $x \times 0 = 0$, and sentence (6) holds. If $y \in N'$, then, for some $z \in R$, $x \cdot y = z \cdot t$ and $x \cdot Sy = x \cdot (y + t) = (z \cdot t) + (x \cdot t) = (z + x) \cdot t$. Consequently, $x \times y = z$ and $x \times Sy = z + x = (x \times y) + x$, so that (7) holds.

Finally, let $x, y \in N'$. As before, for some $z \in R$, $x \cdot y = z \cdot t$ and hence $x \times y = z$. Since x and y are non-negative and t is positive, z must be non-negative. Since $x \cdot t = t \cdot x$ and $y \cdot t = t \cdot y$, we easily obtain $(z \cdot t) \cdot t = (x \cdot y) \cdot t = (x \cdot t) \cdot y = t \cdot (x \cdot y) = (t \cdot z) \cdot t$ and hence $z \cdot t = t \cdot z$. For every $u \in R$ there are $z', z'' \in R$ such that $(u \cdot z) \cdot t = (u \cdot x) \cdot y = (z' \cdot t) \cdot y = (z' \cdot y) \cdot t = (z'' \cdot t) \cdot t$ and therefore $u \cdot z = z'' \cdot t$. Consequently, $x \times y = z \in N'$, and sentence (11) is satisfied.

We have thus shown that the theory T, obtained from the theory T by adjoining some possible definitions to the axiom system, is an extension of an essentially undecidable theory $Q^{(N)}$; hence $Q^{(N)}$ is interpretable in T. Since T is obviously consistent, the conclusion follows at once by Theorem 7(i) of I.4.

By analyzing the proof just given, we easily notice that, if we were exclusively interested in non-densely ordered rings with unit, our argument could be essentially simplified.

III

UNDECIDABILITY OF THE ELEMENTARY THEORY OF GROUPS

BY

ALFRED TARSKI

III

UNDECIDABILITY OF THE ELEMENTARY THEORY OF GROUPS

We consider in this paper an axiomatic theory G with standard formalization, called the *elementary theory of groups* and characterized by the following stipulations: the only non-logical constant of G is the binary operation symbol $\circ$; the set of non-logical axioms of G consists of the three sentences:

$\Gamma_1.$ $x \circ (y \circ z) = (x \circ y) \circ z.$

$\Gamma_2.$ $\vee z(x = y \circ z).$

$\Gamma_3.$ $\vee y(x = y \circ z).$

Hence, a sentence is valid in G if and only if it is satisfied in every system $\langle G, \circ \rangle$ which is a group in the sense of modern algebra (see I.2). The purpose of the paper is to show that the *elementary theory of groups is undecidable.* [1]

Let J be the system of the arithmetic of integers discussed in II.6. Thus, J is a theory with standard formalization; its only non-logical constants are two binary operation symbols, $+$ and $\cdot$; a sentence is valid in J if and only if it is satisfied in the system $\langle I, +, \cdot \rangle$ where I is the set of all integers while $+$ and $\cdot$ are the ordinary arithmetical operations of addition and multiplication.

Let J* and J+ be two other systems of the arithmetic of integers which differ from J in the choice of non-logical constants, but for

---

[1] The main result of this article was stated in the author's talk to the Princeton University Bicentennial Conference on Problems of Mathematics in 1946, and discussed in his talk to a meeting of the Association for Symbolic Logic in 1948. The result is mentioned in [27], p. 763, and a short outline of the proof is given in [39]. (The formulation of the result in [39] is defective; cf. I, footnote 24.) A weaker result in the same direction was announced by Jaśkowski in [11].

which the notion of validity is defined in an entirely analogous way. The non-logical constants of J* are $+$ and the unary operation symbol **K** used to denote the arithmetical operation of forming a square. The non-logical constants of J$^+$ are $+$, the individual constant 1 used with the ordinary arithmetical meaning, and the binary predicate | denoting the relation of divisibility between integers.

THEOREM 1. *Theory* J *is interpretable in Theory* J$^+$.

PROOF: We first show that

(1)   J is interpretable in J*.

We observe that $\cdot$ is the only constant of J which does not occur in J*. Hence (by the remarks in I.4), in order to prove (1) it suffices to construct a theory T, with $+$, $\cdot$, and **K** as the only non-logical constants, and a sentence Ψ in T which satisfy the following conditions:

(2) Ψ is a possible definition of $\cdot$ in J*;

(3)   a sentence of T is valid in T if and only if it is derivable from the set of all valid sentences of J* supplemented by the sentence Ψ;

(4)   T is an extension of J.

The theory T will be uniquely determined by specifying Ψ and by taking (3) as the definition of validity in T. We notice that, in the arithmetic of integers, the product $n \cdot p$ can be defined in terms of sum and square as the only integer m such that

$$m + m + n^2 + p^2 = (n + p)^2.$$

Accordingly we choose for Ψ the sentence

(5) $u = y \cdot z \leftrightarrow (u + u) + (Ky + Kz) = K(y + z).$

The right side Φ of the equivalence (5) is clearly a formula of J*, and the sentence

(6) $\vee u\{\Phi(u) \wedge \wedge x[\Phi(x) \rightarrow x = u]\}$

is easily seen to be valid in J*; hence (2) holds. (See I.4 for the exact sense of (2), and II.1 for the meaning of the expressions $\Phi(x)$ and $\Phi(u)$.)

With every sentence Ω of T we can correlate a sentence Ω^* of J* by eliminating the symbol $\cdot$ with the help of (5). Ω is clearly derivable from the set consisting of Ω^* and (5). Moreover, in view of the intuitive content of (5), it seems obvious that, whenever Ω is valid in J, the correlated sentence Ω^* is valid in J*, and hence Ω is valid in T; a formal proof of this statement presents no essential difficulties. [2] Consequently, (4) holds. Thus the theory T has all the desired properties, and (1) has been proved.

In an entirely analogous way we show that

(7)   J* is interpretable in J+.

The only difference in the argument is that Ψ is replaced by a suitably chosen sentence Ψ' which is a possible definition of **K** in J+. We notice that, in the arithmetic of integers, the square n^2 can be characterized as the only integer m such that (i) $m + n$ is a least common multiple of n and $n + 1$ and (ii) $m - n$ is a least common multiple of n and $n - 1$. (Since any two integers have in general two different least common multiples, neither of the conditions (i) and (ii) alone characterizes m unambiguously.) Consequently we choose as Ψ' the sentence

$$x = \mathbf{K}y \leftrightarrow \wedge z(x + y \mid z \leftrightarrow y \mid z \wedge y + 1 \mid z) \wedge$$
$$\wedge x' \wedge y' \wedge z[x' + y = x \wedge y' + 1 = y \rightarrow (x' \mid z \leftrightarrow y \mid z \wedge y' \mid z)].$$

Theorem 1 follows immediately from (1) and (7). [3]

COROLLARY 2. *There exists a subtheory of Theory* J+ *which is essentially undecidable and finitely axiomatizable.*

[2] An exact proof of this statement must be based upon formal definitions of semantical notions which are involved in the definitions of validity in J and J+ (see I, footnote 7). On the other hand, we could eliminate the use of semantical notions from the whole paper. To this end, however, we should have to replace Theories J and J+ by some of their subtheories, with specified finite axiom systems, which are known to be essentially undecidable; cf. the remark following Corollary 2.

[3] The essence of the proof just outlined clearly consists in showing that in the arithmetic of integers multiplication is definable in terms of addition, divisibility, and the number 1. For positive integers, a stronger result was subsequently obtained by Julia Robinson who showed that multiplication is definable in terms of the successor function and divisibility (see [20], pp. 101 f.); this result was extended by R. M. Robinson (in 1950) to arbitrary integers.

PROOF: By Theorem 12 of II.6, there exists an essentially undecidable and finitely axiomatizable subtheory of J. Hence, by Theorem 1 just proved and by Theorem 7 of I.4, we obtain the conclusion at once.

From the proof of Theorem 12 in II.6 it is seen that we can effectively construct an essentially undecidable and finitely axiomatizable subtheory K of J by specifying its axioms. Hence, by analyzing the proof of Theorem 1, we could also effectively construct an essentially undecidable and finitely axiomatizable subtheory K^+ of J^+, and we could use K^+ instead of J^+ in our further discussion.

As outlined in I.5, given any theory T and any unary predicate **P**, we can form the theory $T^{(P)}$ obtained from T by relativizing the quantifiers to **P**; the set of non-logical constants of $T^{(P)}$ consists of **P** and of all non-logical constants of T. In particular, taking Theory J^+ for T and the predicate **I** (denoting the set of all integers) for **P**, we obtain the theory $J^{+(I)}$ in which the set of non-logical constants consists of four symbols: **I**, **+**, **1**, and **|**.

THEOREM 3. *Theory $J^{+(I)}$ is weakly interpretable in some inessential extension of Theory* G.

PROOF: Let G' be the axiomatic theory obtained from G by including the individual constant **c** in the system of non-logical constants, but without changing the system of non-logical axioms. By I.2, G' is an inessential extension of G. By I.4, in order to show that $J^{+(I)}$ is weakly interpretable in G', it suffices to construct a theory T with the following properties:

(1)   T is consistent;
(2)   T is a common extension of G' and $J^{+(I)}$;
(3)   for each of the non-logical constants of $J^{+(I)}$ there is a valid sentence of T which is a possible definition of this constant in G'.

T is constructed as a theory with standard formalization in which the set of non-logical constants consists of six symbols, **o**, **c**, **I**, **+**, **1**, **|**. To define validity in T we consider the system $\mathfrak{M} = \langle G, o, c, I', +', 1', |' \rangle$ characterized as follows: G is the set of all permutations of the set I of all integers, i.e., the set of all biunique functions f for which I is both the domain and the range. Given two functions $f,g \in G$, we understand by $f \circ g$ their composition, i.e., the function h satisfying the condition $h(k) = f(g(k))$ for every $k \in I$. By c we denote the successor function previously denoted by S, i.e., the function determined by the formula

(4) $$c(k) = k + 1 \text{ for every } k \in I.$$

I' is defined to be the set of all iterations (powers) of c, i.e., the set of all functions c^m, with $m \in I$, such that

(5) $$c^m(k) = k + m \text{ for every } k \in I.$$

Clearly, I' is a subset of G. We identify $+'$ with $\circ$, and $1'$ with c. Finally, we denote by $|'$ the relation which holds between any two functions $f,g \in I'$ if and only if g is an iteration of f. A sentence Φ is by definition valid in T if it is satisfied in the system $\mathfrak{M}$ just described.

The theory T is obviously consistent. Clearly, G is a group of transformations; hence, the non-logical axioms of G' are satisfied in $\mathfrak{M}$, and T is an extension of G'. By (5) we have a one-to-one correspondence between arbitrary integers m and the correlated functions $c^m \in I'$, and under this correspondence, for any $m,n \in I$,

(6) $$c^m +' c^n = c^m \circ c^n = c^{m+n},$$

(7) $c^m \mid' c^n$ if and only if $m \mid n$ (i.e., m is a divisor of n),

as well as, by (4),

(8) $$c^1 = c = 1'.$$

Thus the systems $\langle I, +, 1, | \rangle$ and $\langle I', +', 1', |' \rangle$ are isomorphic. Hence, Φ being a valid sentence of J, the relativized sentence $\Phi^{(l)}$ is satisfied in our model $\mathfrak{M}$ and is therefore valid in T. [4] Consequently, T is an extension of $\mathsf{J}^{+(l)}$.

Conditions (1) and (2) have thus been established. To obtain (3), we consider the following four sentences of T:

(9) $lx \leftrightarrow x \circ c = c \circ x,$

---

[4] Cf. footnote 2.

(10) $$x = y + z \leftrightarrow x = y \circ z,$$

(11) $$x = 1 \leftrightarrow x = c,$$

(12) $$x \mid y \leftrightarrow x \circ c = c \circ x \land y \circ c = c \circ y \land$$
$$\land z(x \circ z = z \circ x \rightarrow y \circ z = z \circ y).$$

Sentences (9), (10), (11), (12) are, respectively, possible definitions of the constants I, $+$, 1, $\mid$ in G'. It remains to be shown that these four sentences are valid in T, i.e., satisfied in the model $\mathfrak{M}$. This is obvious as far as (10) and (11) are concerned since we have simply identified (in $\mathfrak{M}$) $+'$ and $1'$ with $\circ$ and c. We can thus restrict ourselves to (9) and (12).

If $f \in I'$, i.e., $f = c^m$ for some $m \in I$, we see from (6) and (8) that $f \circ c = c \circ f$. If, conversely, f is any function in G such that $f \circ c = c \circ f$, we have by (4)

$$f(c(k)) = c(f(k)), \text{ i.e., } f(k + 1) = f(k) + 1$$

for every integer k. Hence by induction

$$f(k + m) = f(k) + m$$

for every non-negative integer m; by replacing in this formula k by $k-m$, we notice that it holds for negative integers as well. In particular, letting $k = 0$, we obtain

$$f(m) = m + f(0)$$

for every integer m. Putting

$$f(0) = n$$

we see from (5) that f coincides with c^n and therefore belongs to I'. Thus,

(13) $$f \in I' \text{ if and only if } f \circ c = c \circ f;$$

in other words, sentence (9) is satisfied in $\mathfrak{M}$.

If

(14) $$f \mid' g,$$

then, by definition, f and g are in I' and hence, by (13),

(15) $$f \circ c = c \circ f \text{ and } g \circ c = c \circ g.$$

Moreover, f and g are of the form

(16) $$f = c^m, \ g = c^n,$$

and therefore, by (7),

(17) $$m \mid n.$$

We let $n = m \cdot p$ and, by an easy induction on non-negative p's (with an obvious extension to negative p's), we conclude from (5), (16), and (17) that

(18) $f \circ h = h \circ f$ implies $g \circ h = h \circ g$ for every function $h \in G$.

If, conversely, conditions (15) and (18) hold, then, by (13), f and g must be of the form (16) for some integers m and n. Assume first that $m \neq 0$ and consider the function h defined by the formulas:

(19) $$h(k) = k + m \text{ if } m \text{ divides } k,$$
(20) $$h(k) = k \text{ if } m \text{ does not divide } k.$$

Clearly, $h \in G$. From (19) and (20) we see that

$$h(k + m) = h(k) + m$$

and hence, by (15),

$$c^m(h(k)) = h(c^m(k))$$

for every integer k. Therefore, by (16),

(21) $$f \circ h = h \circ f$$

and hence, by (18),

(22) $$g \circ h = h \circ g,$$

i.e.,

(23) $$h(k) + n = c^n(h(k)) = h(c^n(k)) = h(k + n)$$

for every k. Putting $k = 0$ in (23) and using (19), we see that $h(n) = m + n$. Thus, by (20), we must have $m \mid n$; hence, by (7) and (16), we obtain (14).

We have still to consider the case $m = 0$. By (5) and (16), f is then the identity function. Therefore we have (21) and hence, by

(18), we also have (22) for *every* function $h \in G$. If we had $n \neq 0$, we could choose an arbitrary integer $m' \neq 0$ which does not divide n, and could take for h the function defined in (19) and (20) with $m = m'$; the argument outlined above (for $m \neq 0$) shows that in this case (22) would fail. Hence, $n = 0$, $f = g$, and (14) again holds.

We have thus proved that, for arbitrary functions $f, g \in G$, formula (14) is equivalent to the conjunction of (15) and (18); in other words, sentence (12) is satisfied in the system $\mathfrak{M}$.

Since all the sentences (9)–(12) are satisfied in our model, condition (3) is established, and the proof of Theorem 3 is complete. [5]

Theorems 1 and 3 imply that Theory J, the arithmetic of integers, is relatively weakly interpretable (in the sense of I.5) in some inessential extension of Theory G, the elementary theory of groups. It would be of interest to know whether this result can be improved by showing that J is weakly interpretable (in the non-relativized sense) in an inessential extension of G. This would amount to showing that there is binary operation on $I \times I$ to I under which the integers form a group and which has the property that, in terms of this operation and some invidual integers, the ordinary arithmetical addition and multiplication can be defined within the first-order predicate logic. In this connection Mostowski has recently pointed out that Theory J is not weakly interpretable in Theory G itself; in fact, there is no binary operation under which the integers form a group and which has the property that arithmetical addition and multiplication can be defined in terms of this operation *alone*. This follows easily from the fact that every group (with more than two elements) has a non-trivial automorphism, while the ring of integers has no such automorphism.

THEOREM 4.   (i) *Theory* G *is undecidable, and the same applies to every subtheory of* G *which has the same constants as* G.

(ii) *There exists a finitely axiomatizable extension of* G *which has the same constants as* G *and which is essentially undecidable.*

---

[5] Another proof of Theorem 3 (based upon a different construction of T) is briefly outlined in [39].

PROOF: By Corollary 2, there is a subtheory T of J^+ which is finitely axiomatizable and essentially undecidable. Therefore, by Theorems 9 and 10 of I.5, the correlated subtheory $\mathsf{T}^{(l)}$ of $\mathsf{J}^{+(l)}$ is also finitely axiomatizable and essentially undecidable. By Theorem 3, $\mathsf{T}^{(l)}$ is weakly interpretable in an inessential extension of G. Hence, by Theorem 8 of I.4, we immediately obtain both conclusions.

The elementary theory of groups, though undecidable, is not essentially undecidable. If, e.g., we include the commutative law

Γ_4. $x \circ y = y \circ x$

in the system of axioms of G, we obtain an extension of G, in fact, the elementary theory of Abelian groups, which is known to be decidable (a result of Wanda Szmielew; see [26] and [27]). By analyzing the argument in Theorem 3 we see that our whole proof of the undecidability of G depends essentially on the fact that the commutative law is not valid in G.

By Theorem 4(ii), there is an extension of G, i.e., the elementary theory of a particular class of groups, which is finitely axiomatizable and essentially undecidable. However, no simple and mathematically interesting theory of this kind has so far been actually constructed. On the other hand, various interesting extensions of G are known to be undecidable, though not essentially undecidable. This applies, for instance, to the *elementary theory of centerless groups*, i.e., to the theory C obtained from G by including the following sentence in the system of non-logical axioms:

Γ_5. $\wedge y(x \circ y = y \circ x) \rightarrow \wedge y(x \circ y = y)$.

In fact, the proof of undecidability for C is practically the same as for G; from the proof of Theorem 3 it follows immediately that the group of transformations G involved in the argument is centerless.

For many extensions of G, e.g., for the elementary theories of finite groups and non-Abelian free groups, the decision problem remains open. No general and workable criteria, of mathematical or metamathematical nature, are known which would permit us to decide, for a great variety of special groups and special classes of groups, whether or not their elementary theories are decidable.

By our main result, the set of all valid sentences of G is not recursive. The problem remains open whether the same applies to the subset of this set consisting of all so-called open, or universal, sentences (which are valid in G), i.e., sentences of the form

$$\wedge x_1 \ldots \wedge x_n \Phi$$

where Φ is a formula containing no quantifiers. This problem is nothing else but an equivalent formulation of the well-known word problem for groups. [6]

Among subtheories of G which are undecidable by Theorem 4 we may mention the *elementary theory of groupoids (associative systems)*, i.e., the theory obtained from G by omitting Axioms Γ_2 and Γ_3, and the *elementary theory of semigroups*, i.e., the theory obtained from G by replacing Γ_2 and Γ_3 by the two cancellation laws:

$\Gamma_6.$ $\qquad\qquad x \circ y = x \circ z \to y = z,$

$\Gamma_7.$ $\qquad\qquad x \circ z = y \circ z \to x = y;$

and possibly also by the following law which states the existence of a unit element;

$\Gamma_8.$ $\qquad\qquad \vee x \wedge y(x \circ y = y \wedge y \circ x = y).$

Some unpublished results jointly obtained by Wanda Szmielew and the author enable us to exhibit interesting extensions of the elementary theories of groupoids and semigroups which are finitely axiomatizable and essentially undecidable. Consider, for instance, the axiomatic theory F in which $\circ$ occurs as the only non-logical constant while the set of non-logical axioms consists of Γ_1, Γ_6, Γ_7, and the following sentence:

$\Gamma_9.$ $\quad \vee a \vee b\{\wedge x \wedge y(x \circ a \neq y \circ b) \wedge$

$\qquad\qquad \wedge z[z = a \vee z = b \vee \vee x(z = x \circ a) \vee \vee y(z = y \circ b)]\}.$

---

[6] Concerning the equivalence of our formulation of the word problem with the usual one see [14], in particular p. 68.

F is a finitely axiomatizable fragment of the elementary theory of free semigroups with two generators. It can be shown that Theory Q, i.e., the essentially undecidable fragment of arithmetic discussed in II, is relatively interpretable in some inessential extension of F and that consequently, by Theorems 7(i) and 10 of I, F is essentially undecidable. Some subtheories of F are known which are also essentially undecidable, e.g., the subtheory obtained by omitting Γ_6 in the axiom system of F. In an analogous way we can construct finitely axiomatizable and essentially undecidable fragments of the elementary theory of free semigroups with n generators, for any given natural number $n \geqslant 2$. [7]

---

[7]   Consequently, the elementary theory of free semigroups with n generators is itself undecidable and, in fact, essentially undecidable. This result, however, is not new since it follows directly from the discussion in [18].

BIBLIOGRAPHY

[1] PAUL BERNAYS. *A system of axiomatic set theory.* Part I, Journal of Symbolic Logic, vol. 2 (1937), pp. 65–77; Part II, ibid., vol. 6 (1941), pp. 1–17; Part III, ibid., vol. 7 (1942), pp. 65–89; Part IV, ibid., vol. 7 (1942), pp. 133–145; Part V, ibid., vol. 8 (1943), pp. 89–106.

[2] ALONZO CHURCH. *A note on the Entscheidungsproblem.* Journal of Symbolic Logic, vol. I (1936), pp. 40–41; *Correction,* ibid., pp. 101–102.

[3] ALONZO CHURCH. *An unsolvable problem of elementary number theory.* Americal Journal of Mathematics, vol. 58 (1936), pp. 345–363.

[4] ALONZO CHURCH. *Special cases of the decision problem.* Revue Philosophique de Louvain, vol. 49 (1951), pp. 203–221.

[5] KURT GÖDEL. *Die Vollständigkeit der Axiome des logischen Funktionenkalküls.* Monatshefte für Mathematik und Physik, vol. 37 (1930), pp. 349–360.

[6] KURT GÖDEL. *Über die Länge der Beweise.* Ergebnisse eines Mathematischen Kolloquiums, vol. 7 (1936), pp. 23–24.

[7] KURT GÖDEL. *Über formal unentscheidbare Sätze der Principia Mathematica und verwandter Systeme I.* Monatshefte für Mathematik und Physik, vol. 38 (1931), pp. 173–198.

[8] ANDRZEJ GRZEGORCZYK. *Undecidability of some topological theories.* Fundamenta Mathematicae, vol. 38 (1951), pp. 137–152.

[9] DAVID HILBERT and PAUL BERNAYS. *Grundlagen der Mathematik.* Vol. 1, Berlin 1934, XII + 471 pp.; vol. 2, Berlin 1939, XII + 498 pp.

[10] ANTONI JANICZAK. *A remark concerning decidability of complete theories.* Journal of Symbolic Logic, vol. 15 (1950), pp. 277–279.

[11] STANISŁAW JAŚKOWSKI. *Sur le problème de decision de la topologie et de la théorie des groupes.* Colloquium Mathematicum, vol. 1 (1947–1948), pp. 176–178.

[12] LÁSZLÓ KALMÁR. *Zum Entscheidungsproblem der mathematischen Logik.* Verhandlungen des Internationalen Mathematiker-Kongresses (Zürich 1932), vol. 2, pp. 337–338.

[13] S. C. KLEENE. *Recursive predicates and quantifiers*. Transactions of the American Mathematical Society, vol. 53 (1943), pp. 41–73.

[14] J. C. C. MCKINSEY. *The decision problem for some classes of sentences without quantifiers*. Journal of Symbolic Logic, vol. 8 (1943), pp. 61–76.

[15] ANDRZEJ MOSTOWSKI. *Sentences undecidable in formalized arithmetic*. Amsterdam 1952, VIII+117 pp.

[16] ANDRZEJ MOSTOWSKI and ALFRED TARSKI. *Undecidability in the arithmetic of integers and in the theory of rings*. Journal of Symbolic Logic, vol. 14 (1949), p. 76.

[17] M. PRESBURGER. *Über die Vollständigkeit eines gewissen Systems der Arithmetik ganzer Zahlen, in welchem die Addition als einzige Operation hervortritt*. Comptes-rendus du I Congrès des Mathématiciens des Pays Slaves (Warszawa 1929), pp. 92–101 and 395.

[18] WILLARD VAN ORMAN QUINE. *Concatenation as a basis for arithmetic*. Journal of Symbolic Logic, vol. 11 (1946), pp. 105–114.

[19] WILLARD VAN ORMAN QUINE. *Mathematical logic*. New York 1940, XIV+348 pp.

[20] JULIA ROBINSON. *Definability and decision problems in arithmetic*. Journal of Symbolic Logic, vol. 14 (1949), pp. 98–114.

[21] JULIA ROBINSON. *General recursive functions*. Proceedings of the American Mathematical Society, vol. 1 (1950), pp. 703–718.

[22] RAPHAEL M. ROBINSON. *An essentially undecidable axiom system*. Proceedings of the International Congress of Mathematicians (Cambridge 1950), vol. 1, pp. 729–730.

[23] RAPHAEL M. ROBINSON. *Undecidable rings*. Transactions of the American Mathematical Society, vol. 70 (1951), pp. 137–159.

[24] BARKLEY ROSSER. *Extensions of some theorems of Gödel and Church*. Journal of Symbolic Logic, vol. 1 (1936), pp. 87–91.

[25] C. RYLL-NARDZEWSKI. *The role of the axiom of induction in elementary arithmetic*. Fundamenta Mathematicae, vol. 39 (1952), pp. 239–263.

[26] WANDA SZMIELEW. *Arithmetical properties of Abelian groups*. Doctoral dissertation, University of California, Berkeley 1950.

[27] WANDA SZMIELEW. *Decision problem in group theory*. Proceedings of the Tenth International Congress of Philosophy (Amsterdam 1948), fasc. 2, pp. 763–766.

[28] WANDA SZMIELEW and ALFRED TARSKI. *Mutual interpretability of some essentially undecidable theories.* Proceedings of the International Congress of Mathematicians (Cambridge 1950), vol. 1, p. 734.

[29] WANDA SZMIELEW and ALFRED TARSKI. *Theorems common to all complete and axiomatizable· theories.* Bulletin of the American Mathematical Society, vol. 55 (1949), p. 1075.

[30] ALFRED TARSKI. *A decision method for elementary algebra and geometry.* Second edition, Berkeley and Los Angeles 1951, VI + 63 pp.

[31] ALFRED TARSKI. *Der Wahrheitsbegriff in den formalisierten Sprachen.* Studia Philosophica, vol. 1 (1936), pp. 261–405.

[32] ALFRED TARSKI. *Fundamentale Begriffe der Methodologie der deduktiven Wissenschaften I.* Monatshefte für Mathematik und Physik, vol. 37 (1930), pp. 361–404.

[33] ALFRED TARSKI. *Grundzüge des Systemenkalküls.* Part I, Fundamenta Mathematicae, vol. 25 (1935), pp. 503–526.

[34] ALFRED TARSKI. *On essential undecidability.* Journal of Symbolic Logic, vol. 14 (1949), pp. 75–76.

[35] ALFRED TARSKI. *Remarks on the formalization of the predicate calculus.* Bulletin of the American Mathematical Society, vol. 57 (1951), pp. 81–82.

[36] ALFRED TARSKI. *Sur les ensembles définissables de nombres réels I.* Fundamenta Mathematicae, vol. 17 (1931), pp. 210–239.

[37] ALFRED TARSKI. *Two general theorems on undefinability and undecidability.* Bulletin of the American Mathematical Society, vol. 59 (1953) pp. 365–366.

[38] ALFRED TARSKI. *Über den Begriff der logischen Folgerung.* Actes du Congrès International de Philosophie Scientifique (Paris 1935), vol. 7, pp. 1–11.

[39] ALFRED TARSKI. *Undecidability of group theory.* Journal of Symbolic Logic, vol. 14 (1949), pp. 76–77.

[40] ALFRED TARSKI. *Undecidability of the theories of lattices and projective geometries.* Journal of Symbolic Logic, vol. 14 (1949), pp. 77–78.

INDEX

(i) To avoid repetitions, words and successions of words are often replaced by dashes, each dash replacing one word.

(ii) Synonyms of a term and symbols for a term, insofar as they are actually used in the book, are given in brackets. Specific contexts in which a term occurs are indicated in parentheses.

(iii) Numbers following the name of an author indicate those pages on which either the author himself is mentioned or a work of his is referred to.

CPSIA information can be obtained at www.ICGtesting.com
Printed in the USA
LVOW04s0455070115

421736LV00026B/581/P